NATIONAL ACADEMIES *Sciences Engineering Medicine*

NATIONAL ACADEMIES PRESS
Washington, DC

Corrosion of Buried Steel at New and In-Service Infrastructure

Committee on the Corrosion of Buried Steel at New and In-Service Infrastructure

Committee on Geological and Geotechnical Engineering

Board on Earth Sciences and Resources

Division on Earth and Life Studies

National Materials and Manufacturing Board

Division on Engineering and Physical Sciences

Transportation Research Board

Consensus Study Report

NATIONAL ACADEMIES PRESS 500 Fifth Street NW Washington, DC 20001

This activity was supported by contracts between the National Academy of Sciences and the American Society of Civil Engineers, Association for Mechanically Stabilized Earth, Federal Highway Administration, International Association of Foundation Drilling, National Science Foundation, and U.S. Army Corps of Engineers. Any opinions, findings, conclusions, or recommendations expressed in this publication do not necessarily reflect the views of any organization or agency that provided support for the project.

International Standard Book Number-13: 978-0-309-69267-0
International Standard Book Number-10: 0-309-69267-9
Digital Object Identifier: https://doi.org/10.17226/26686

This publication is available from the National Academies Press, 500 Fifth Street, NW, Keck 360, Washington, DC 20001; (800) 624-6242 or (202) 334-3313; http://www.nap.edu.

Suggested citation: National Academies of Sciences, Engineering, and Medicine. 2023. *Corrosion of Buried Steel at New and In-Service Infrastructure*. Washington, DC: The National Academies Press. https://doi.org/10.17226/26686.

The **National Academy of Sciences** was established in 1863 by an Act of Congress, signed by President Lincoln, as a private, nongovernmental institution to advise the nation on issues related to science and technology. Members are elected by their peers for outstanding contributions to research. Dr. Marcia McNutt is president.

The **National Academy of Engineering** was established in 1964 under the charter of the National Academy of Sciences to bring the practices of engineering to advising the nation. Members are elected by their peers for extraordinary contributions to engineering. Dr. John L. Anderson is president.

The **National Academy of Medicine** (formerly the Institute of Medicine) was established in 1970 under the charter of the National Academy of Sciences to advise the nation on medical and health issues. Members are elected by their peers for distinguished contributions to medicine and health. Dr. Victor J. Dzau is president.

The three Academies work together as the **National Academies of Sciences, Engineering, and Medicine** to provide independent, objective analysis and advice to the nation and conduct other activities to solve complex problems and inform public policy decisions. The National Academies also encourage education and research, recognize outstanding contributions to knowledge, and increase public understanding in matters of science, engineering, and medicine.

Learn more about the National Academies of Sciences, Engineering, and Medicine at **www.nationalacademies.org**.

---

[1] Resigned June 2021.

Acknowledgments

This Consensus Study Report was reviewed in draft form by individuals chosen for their diverse perspectives and technical expertise. The purpose of this independent review is to provide candid and critical comments that will assist the National Academies of Sciences, Engineering, and Medicine in making each published report as sound as possible and to ensure that it meets the institutional standards for quality, objectivity, evidence, and responsiveness to the charge. The review comments and draft manuscript remain confidential to protect the integrity of the process.

We thank the following individuals for their review of this report:

JAMES BRENNAN, Kansas Department of Transportation
JERRY A. DiMAGGIO, Applied Research Associates
JAMES ELLOR, Elzly Technology Corporation
KHALID FARRAG, Gas Technology Institute
ROBERT E. MELCHERS, The University of Newcastle
JAMES K. MITCHELL (NAS/NAE), Virginia Polytechnic Institute and State University
CHRISTOPHER RAMSAY, Missouri University of Science and Technology
JOHN SENKO, The University of Akron
NARASI SRIDHAR, The Ohio State University
SEBASTIAN UHLEMANN, Lawrence Berkeley National Laboratory

Although the reviewers have provided many constructive comments and suggestions, they were not asked to endorse the conclusions or recommendations nor did they see the final draft of the report before its release. The review of this report was overseen by **GEORGE M. HORNBERGER**, Vanderbilt University, and **CLYDE L. BRIANT**, Brown University. Appointed by the National Academies, they were responsible for making certain that an independent examination of this report was carried out in accordance with institutional procedures and that all review comments were carefully considered. Responsibility for the final content of this report rests entirely with the authoring committee and the institution.

The committee benefited from the input of numerous individuals who gave presentations during its open sessions and participated in its information-gathering workshop. Appendix B provides the names of meeting presenters and workshop participants.

Contents

Summary

Steel is a common component of the nation's infrastructure, and steel corrosion costs the United States an estimated 3–4 percent of its gross domestic product annually. Infrastructure failure due to steel corrosion can result in loss of life, destruction of property, damage to the environment, disruption of energy transport, and major economic losses. Steel is composed of iron mixed with metallic and nonmetallic elements (e.g., manganese and carbon, respectively) in various amounts to create alloys for different purposes. It is often buried in the earth as a primary construction material (as in fuel or water pipelines) or as a support element (as in structural foundations or retaining structures). When buried, steel is often in direct contact with soil (i.e., unconsolidated earth materials composed of inorganic solids, liquids, and gases), rock, engineered fills, and grouts and can be exposed to complex and corrosive environments that are difficult to characterize.

The corrosion of steel occurs predominantly through an electrochemical process involving loss of electrons at one site on the steel (the anodic site where oxidation occurs) and gain of electrons at another site on the steel (a cathodic site where reduction occurs). An electrolyte in contact with both sites conducts ions between anode and cathode. In the case of buried steel, the soil, rock, or grout with which the steel is in contact is the electrolyte. Properties such as moisture content, grain size and mineralogy, compaction, pore space, pH, soluble salt content, sulfide content, dissolved oxygen content, temperature, organic content, microbiology, redox potential, and electrical properties such as resistivity are either conducive to or indicative of the corrosivity of the electrolyte. The mechanisms of steel corrosion are well understood theoretically, but the complexity of the subsurface makes it difficult to predict with certainty where, when, by which mechanisms, and at what rates corrosion will occur, and therefore to determine the most effective infrastructure design and corrosion protection measures. When corrosion occurs, it often is difficult to determine whether it is the result of inaccurate site characterization and prediction of environmental corrosivity, poor choice of steel design or protection, material defects, poor construction or quality control practices, insufficient monitoring or maintenance, changes in subsurface conditions, or some combination of these factors.

There are two general approaches to protect buried steel against corrosion, although not all applications fit cleanly into these approaches. The first is the "corrosion allowance" approach in which steel is designed with extra cross-sectional thickness to compensate for expected loss of steel to corrosion over a designated performance period. This is the practice applied most often in structural foundations, earth retaining structures, dams, and tunnels. This report refers to the industries that apply this approach as the "geo-civil industries." The geo-civil industries attempt to characterize the corrosivity of the subsurface and design the infrastructure accordingly;

once buried, steel undergoes little monitoring. The second approach is a "corrosion avoidance" approach, which is applied, for example, by the oil and gas pipeline industries. Corrosion is prevented or minimized through such means as coatings and cathodic protection (CP). For practical reasons, the oil and gas pipeline industries rely less on site characterization (pipelines can be hundreds of kilometers long) and more on monitoring the effectiveness of the protection measures. Regardless of the corrosion protection approach applied, the same corrosion mechanisms impact the steel.

THE STUDY CHARGE

The American Society of Civil Engineers, Association for Mechanically Stabilized Earth, Federal Highway Administration, International Association of Foundation Drilling, National Science Foundation, U.S. Army Corps of Engineers, and the presidents of the National Academies of Sciences, Engineering, and Medicine (the National Academies) requested an interdisciplinary ad hoc committee of the National Academies to summarize mechanisms for corrosion of steel within earth materials. The committee was asked to assess current practices for characterizing the subsurface environment for corrosivity and to assess the use, efficacy, and uncertainties of methods used to predict, identify, and monitor corrosion of steel in earth materials at existing and new infrastructure. The committee organized a workshop of experts to investigate field, laboratory, and modeling methods for characterizing corrosion of steel buried in earth materials and new developments in prediction and monitoring corrosion in earth applications and environments. This report describes the findings of the committee, identifies gaps in knowledge, and recommends the research needed to improve the long-term performance of steel in earth applications. The report focuses on steel buried in and in direct contact with soil and discusses generalized practices of the geo-civil and oil and gas industries. The report does not provide "best practices" or discuss corrosion of steel in concrete (except reference to cementitious grouts), steel in marine environments, or interior pipeline corrosion. Stainless steels are mentioned only briefly.

Early in its deliberations, the multidisciplinary committee recognized little common language use surrounding key issues related to corrosion, corrosion rates, corrosivity, and corrosion protection. The committee learned that important terms are defined differently across sectors and even within sectors. Practices and guidance vary greatly between sectors, and knowledge transfer does not occur often or easily across sectors. As a result, the committee found it necessary, at a primer level, to identify and define relevant types of steel, the fundamental mechanisms of steel corrosion, the different subsurface conditions that affect corrosivity, and the range of buried-steel applications and protection measures to develop a common basis of understanding. The first chapters of this report provide similar primer-level material and provide definitions of terms that are inconsistent across industries.

OVERARCHING CONCLUSIONS

Site characterization and infrastructure design and modeling practices are often driven by standards used for efficiency and reproducibility but that were likely derived from practices developed for different purposes and then repurposed for convenience. Many of those standards are not informed by empirically based knowledge about how given conditions affect corrosivity, and they generally do not consider how the combined effects of subsurface physical, chemical, and microbiological properties of the complex subsurface environment affect corrosivity. Researchers in industry and academe often rely on limited corrosion-related data that were collected and analyzed in the mid-twentieth century. Compounding the uncertainties already inherent in the methodologies is the inconsistent use of vocabulary between industries—and disciplines within industries—and that data are not routinely shared between industries. The different approaches adopted by the geo-civil and oil and gas pipeline industries have resulted in intellectual and practical silos that have led to parallel—or even competing—research initiatives and foci.

Significantly improving understanding of corrosion mechanisms and rates for buried steel will require multidisciplinary approaches to investigation informed by an inclusive vocabulary that is easily translated among disciplines. Comprehensive long-term multivariate experiments are needed that will allow collection of observational data regarding the individual and combined contributions of subsurface properties on environmental corrosivity, corrosion mechanisms, and corrosion rates. Reliable, accessible, and searchable databases—designed to protect proprietary information—could be established to house data collected through that experimentation. Decision support systems (DSSs) for site characterization program design and risk-informed decision making can be developed to inform future characterization, design, modeling, and infrastructure management decisions. Advanced data analytics should be systematically applied to presently available data to better understand important statistical correlations between properties that affect corrosivity and resulting corrosion rates. Methods such as machine learning could later be utilized as data repositories are developed and populated with robust and complete datasets.

IMPROVED COMMUNICATIONS

A multidisciplinary approach that combines, for example, geotechnical engineering, structural engineering, earth science, material science, hydrology, metallurgy, corrosion engineering and modeling, geophysics, geochemistry, and microbiology is necessary to understand corrosion of steel in subsurface environments. However, there is no common vocabulary for corrosion among those fields. As a result, there can be miscommunication between practitioners that can result in uninformed decisions (e.g., "corrosion potential" is used by geotechnical professionals to refer to the possibility of corrosion, whereas the term has a specific electrochemical meaning in corrosion engineering). A common lexicon with more technically precise terminology will increase the effectiveness of communication and collaboration between disciplines and industries and facilitate work toward common goals: better understanding of the corrosivity of a subsurface environment, better prediction of corrosion and corrosion rates, and more effective design, construction, and management of buried steel infrastructure.

> **Recommendation 1: Standards-making bodies from different industries, in collaboration with the public agencies with responsibilities related to buried steel infrastructure, and researchers interested in understanding and preventing buried steel corrosion should develop a common lexicon with precise definitions associated with corrosion of steel and the characterization and monitoring of subsurface environments in which steel is buried or placed.**

Because standards-making bodies already influence practices in the public and private sectors, their collaborative development and dissemination of a common lexicon would help the technical communities that they serve incorporate the vocabulary into their work. Professional societies (e.g., the Association for Materials Protection and Performance, the American Association of State Highway and Transportation Officials, and ASTM International) might collaborate to develop this lexicon. Agreeing on a lexicon will create opportunities to advance knowledge and innovations across disciplines or move research into practice.

MULTIDISCIPLINARY RESEARCH AND LONGITUDINAL EXPERIMENTATION

Engineers who conduct site characterization investigations are rarely knowledgeable about corrosion mechanisms, and corrosion engineers are often unfamiliar with the complexity and heterogeneity of the soil–groundwater–gas electrolyte that complicates realistic modeling of subsurface corrosion. Few engineers are familiar with, for example, the importance of microbially influenced corrosion (MIC) or of the geophysics that might be used to characterize the subsurface. Standards have been developed and are applied based on, for example, resistivity techniques, but often without an understanding of the significance of the results (e.g., that resistivity is not a measure of soil corrosivity but rather an indicator of possible corrosivity). Multidisciplinary research efforts are required to build the knowledge necessary to advance practice to more accurately predict corrosion and aid management capabilities (e.g., to move from overly simplistic corrosion prediction models to more sophisticated modeling techniques). Investigation and identification of appropriate design and management of steel require the combined specialty knowledge of structural and geotechnical engineers, hydrogeologists, geophysicists, CP specialists, corrosion engineers, microbial testing experts, and those with expertise in chemical analysis of fill and soil materials.

There is a need for public agencies, industry groups, and academe with interest in or responsibilities related to corrosion of buried steel to formalize collaborative efforts to identify and facilitate multidisciplinary research for improved prediction of, protection against, and monitoring of the corrosion of buried steel. Primary goals of such collaborative multidisciplinary efforts include increasing understanding of how multiple ground conditions contribute to corrosivity and corrosion mechanisms and translating fundamental research discoveries into practice. Multidisciplinary research in corrosion science would expose researchers and practitioners to new concepts and provide information to synthesize collective and new knowledge about subsurface corrosivity, corrosion mechanisms, and prediction of corrosion rates. Topics to be explored include the combined effects of different soil properties on corrosivity; better ways to characterize the subsurface by combining geophysical, geochemical, hydrological, and microbiological techniques; and the ground response to a changing climate and its effects on corrosivity (given that increased temperatures will accelerate corrosion-dependent chemical reactions and will change the distribution of microorganisms and their rates of activity). Multidisciplinary teams of experts including geotechnical and structural engineers, metallurgists, materials scientists, hydrologists, geochemists, geophysicists, microbiologists, and others will need to collaboratively design and implement the necessary research.

Data from statistically sound, long-term multivariate experiments that involve observations from steel buried in the subsurface make quantifying the fundamental relationships that control corrosion rates possible. Uncertainties in the conclusions drawn from analysis of data in existing databases are considerable because soil environments related to those data were not thoroughly characterized, burial depths and exposure times varied, climate conditions were reported as averages, many soil properties were measured off-site, and the statistical designs associated with the original experiments were weak. Needed is controlled longitudinal research that quantifies the conditions that lead to increased corrosivity, that identify corrosion mechanisms under those conditions, and that allow prediction of corrosion rates.

> **Recommendation 2: Coordinated groups of multidisciplinary researchers, supported through commitments from private- and public-sector organizations and agencies with interest in or responsibilities related to buried steel infrastructure, should conduct comprehensive, long-term experiments to quantify corrosion rates and mechanisms associated with multiple variables on steel buried both in controlled and in carefully characterized natural subsurface conditions.**

Two general mechanisms could facilitate such multidisciplinary research: (1) the organization of formal partnerships (e.g., between industry and academe, between private- and public-sector entities, and between government agencies and academic research facilities); and (2) multidisciplinary research centers that can invest specifically in research that can be scaled up to technologies applicable in practice. Both models could provide opportunities to engage practitioners in research, expose researchers to "real-life" problems, and enhance educational and professional networking opportunities for students.

Experimental results could contribute to a reliable reference database useful to (1) identify the most relevant properties of the subsurface for corrosion rates, (2) quantify the synergistic effects of subsurface properties, (3) assess current corrosion-rate predictive models, and (4) develop corrosion models with less uncertainty in their predictive capabilities. Experiments need to allow multiple observations in the first 5 years to capture changes in corrosion rates when they are greatest, and observations 25 years or longer to capture how those rates attenuate over the service life of infrastructure. The investigations should include laboratory-based experiments with controlled initial and boundary conditions and field-based experiments with extensive soil and hydrologic characterization and monitoring. Studies on buried plain carbon steel are needed, as are experiments on galvanized, aluminized, and polymer coatings, of which the long-term behaviors when buried are not well known.

The experiments should be designed to extract the influences of physical and chemical soil properties, soil water and gaseous phases, and soil spatial variations, and to capture the soil microbiology. Different approaches applied during the same experiments—for example, monitoring electrochemical testing and exhuming coupons for destructive testing—would benefit comparisons and integration. There should also be experiments that document the effects of climate change on corrosivity and corrosion of buried steel infrastructure. Properties relating to the risk of MIC should also be quantified and monitored throughout the study. Robust results from long-term experiments will enable designers, owners, operators, and managers to focus resources on assessing and monitoring the spatial and temporal variations of those properties with the largest impact on corrosivity and corrosion rates at a given site, allowing more efficient design, construction, and management of safer and more resilient infrastructure.

DATA ANALYTICS

Existing characterization approaches do not attempt to weight the corrosivity-inducing effects of all relevant subsurface properties, and they do not describe comprehensively the synergies between subsurface properties. Data analytical techniques (e.g., cluster analysis or Bayesian theory) can be applied to currently available datasets and to new longitudinal experiment data to investigate relationships among properties, and between properties and corrosion rates. Until data from longitudinal and multivariate experiments are available, systematic examination of existing data may be useful to identify statistically important relations among various properties and with corrosion and corrosion rates. Advanced analytical techniques can be applied to identify previously unrecognized relevant synergies between different subsurface properties and their relationships to corrosivity.

> **Recommendation 3: Researchers should use advanced data science techniques on available and new data to determine systematically the statistically important contributions of individual and combined subsurface properties to corrosivity in different subsurface environments.**

Thousands of published papers describe failures of steel infrastructure and laboratory and field testing conducted under varying conditions. The collection of independent observations has not improved predictive capability. Experimental studies that have measured corrosion on buried steel across a number of climates have not been systematically "mined" for data to assess the relationship between, for example, temperature and corrosion and corrosion rates, or for their relationships to MIC. In addition to understanding the relationship of properties and corrosivity, data analytical techniques may help to identify relevant subsurface properties that are not traditionally used to characterize corrosivity. The collective data for MIC, for example, are extensive and could be systematically examined for relationships directly related to MIC (e.g., assimilable nutrients, relationships between electron donors or electron acceptors and aggressive or inhibiting anions). The results will suggest integrated approaches to predict MIC based on the total environment and not just the identification of specific putative microorganisms. Improved estimates of corrosion rates will result from analytical approaches that (1) consider all relevant subsurface properties, (2) apply data-driven weighting factors to relevant subsurface properties, and (3) calculate the synergies between the relevant subsurface properties.

DECISION SUPPORT SYSTEMS

The technical communities and organizations with interest in or responsibilities associated with corrosion of buried steel lack a framework that can tie available multivariate information and guide prioritization of actions using risk-informed approaches. Simplified and empirical methods for modeling metal loss, corrosion rates, and performance of protection systems have limitations and are only applicable for particular sets of conditions. They do not assist in prioritizing actions and investments based on the likelihood and severity of negative impacts (i.e., risk-based decision making). Without a common database of reported case histories of failures, it is difficult to validate models and assess past performance of given protection systems in given environments.

Few protocols in any industry guide proper data collection for characterizing corrosivity and corrosion modeling, and decisions based on those protocols do not benefit from the knowledge or innovations from other sectors. A DSS is a tool that guides decision makers through alternatives. DSSs can be two-dimensional flowcharts or complex digital systems connected to multiple-input databases leading decision makers through numerous options.[1] A robust DSS will help reduce uncertainty in decision making by formalizing standard practices and presenting

---

[1] GeoTechTools (geotechtools.org) is an example of a DSS developed by the Strategic Highway Research Program 2 of the National Academies of Sciences, Engineering, and Medicine, deployed by the Federal Highway Administration, and is now hosted by the Geo-Institute of the American Society of Civil Engineers.

logical and reproducible sequences of decisions based on existing data. They could be particularly helpful in deci-sion areas that rely on large volumes of data combined with predictive modeling. Many existing simple DSSs guide the choice of just a few standard tests from specific industry groups based on a few observed site conditions. A more comprehensive characterization framework and DSS is needed that informs decisions related to subsurface characterization appropriate for multiple combinations of subsurface properties and informed by standards from multiple sectors and industry groups.

> **Recommendation 4: Standards-setting bodies should collaborate with state and federal agencies, indus-try, and academe to create and maintain two decision support systems (DSSs):**
>
> **(1)** **a DSS that guides site characterization and allows selection from among a comprehensive set of characterization tests that are appropriate for temporally and spatially variable surface and subsurface conditions; and**
> **(2)** **a DSS that uses risk-informed decision making to guide corrosion management practices.**

A site characterization DSS will guide characterization design to capture how lateral, vertical, and temporal subsurface variations affect the individual and combined site-specific subsurface characteristics that control cor-rosivity. Uncertainties in characterization results will be captured, and guidance regarding errors in modeling given disparities between measurement scales and the scales at which corrosion initiates on the steel surface will be provided. A DSS for practitioners should outline the minimum field- and laboratory-based information needed to design a site characterization program. Guidance related to spatial and temporal sampling frequencies given the natural setting of the site, land use, infrastructure life cycle, surface and groundwater hydrology, and atmo-spheric conditions will be provided based on the combined knowledge, tools, and standards of multiple industries. The DSS should then inform decisions regarding additional characterization necessary to reduce uncertainties to acceptable levels for modeling.

To make development of this DSS a practical exercise, initial focus should be on subsurface properties most commonly utilized for characterization of corrosivity (e.g., moisture content, resistivity, pH, chlorides, and sulfates), and then expand to promising technologies and less commonly measured properties (e.g., sulfides, microbial-related properties, and redox potential). The system should include both laboratory- and field-based methods and should distinguish which laboratory tests are intended to replicate field conditions versus those that do not. The DSS should continually evolve as understanding of the multivariate controls on corrosivity increases.

As the characterization DSS is developed, a second DSS should be developed based on risk-informed decision making regarding management actions and investments related to corrosion. This second DSS should be developed and maintained in parallel or in concert with the characterization DSS so that it can be informed by outputs from the characterization DSS (including present and future uncertainties about the environment in which the steel is buried). The DSS could assist decisions regarding choice of model to identify the choice of a protection system or the amount of steel needed to compensate for steel loss for a particular site, the depth of burial, specific design details, and other factors. Corrosion management decisions regarding design and modeling for new infrastructure, and the modeling and monitoring of existing infrastructure, could also benefit from the DSS.

The DSS should be interactive and guide decisions based on selected inputs. It should present a compre-hensive list of monitoring techniques using names and descriptions drawn from a common lexicon and should guide decisions regarding monitoring techniques depending on a number of inputs, including the variable to be monitored (coating defects, corrosion rate, CP effectiveness), the depth and dimensions of the infrastructure, and the risk associated with failure. This will require coordinated input, planning, and action of all agencies and organizations with interest in or responsibilities associated with corrosion of buried steel. These groups will need to develop a framework that is able to tie available multivariate information and guide prioritization of actions using risk-informed approaches. The DSS would support better decision making by small companies (e.g., small water utilities) that may not have expertise in all areas and would assist the geo-civil industries in improving asset management.

INDIRECT OBSERVATION

It is not feasible to monitor hundreds of kilometers of pipelines or to repeatedly expose geo-civil steel infrastructure for direct monitoring. It is also not feasible or possible to quantify all the properties relevant to corrosivity for entire infrastructure systems, and certainly not continuously for the infrastructure performance period. Indirect observation through surface monitoring and opportunistic data collection could inform where localized site-specific monitoring is warranted, and results could be used to build a database to inform future research and infrastructure-related decision making.

Recommendation 5: Private- and public-sector infrastructure owners should monitor the land surface for changes that could alter subsurface corrosivity and determine whether localized monitoring of subsurface properties is warranted to maintain infrastructure performance and safety. Surface changes to be monitored include but are not limited to changes in land, land use, and atmospheric conditions that affect surface and groundwater flow, and any asset management decisions by colocated infrastructure managers that might affect subsurface hydrology, geochemistry, microbiology, or the production of stray currents.

Until corrosivity, corrosion, and corrosion rates can be directly measured, infrastructure managers will rely on indirect measurements to estimate corrosivity and corrosion rates. Because changes on the land surface can affect surface and groundwater flow, permeability, soil saturation, soil and water chemistries, subsurface temperatures, and other characteristics that affect corrosivity, monitoring surface changes is a cost-effective early indicator of possible detrimental subsurface change and could indicate where direct measurements are appropriate.

Surface monitoring should include monitoring changes in

- Land use (including upgradient), installation of pavements, large foundations, or other underground structures, and installation of surface, subsurface, or aerial transmission or pipelines that may produce stray currents;
- Land cover, such as transition from rural to urban, and changes in vegetation, including those that may indicate signs of changes in moisture content;
- Installation of upgradient power plants, mining operations, or waste disposal operations that could affect groundwater and soil geochemistries;
- Surface or groundwater flow or retention, changes at the surface that would alter surface-water flow or retention, or the appearance or disappearance of springs;
- Infrastructure conditions such as those that result in the intentional or unintentional release of fluids or that change subsurface temperatures;
- Infrastructure or land management decisions such as use or change in deicing salts on pavements or the application of fertilizers that could leach into the subsurface and increase electrochemical potential or encourage the growth of microbes that influence corrosion;
- Changes in construction practices; and
- Climate and atmospheric conditions (seasonal and global) that result in changes in temperature and precipitation that in turn affect subsurface temperature, groundwater and the groundwater table, degree of saturation, and saltwater intrusion and that may alter the magnitude or frequency of extreme events.

Some surface changes can be monitored from a desktop computer with few computational resources (e.g., using publicly available data to track precipitation and surface temperatures, land use changes, traffic pattern changes, and changes in topography using airborne and satellite data). Some can be monitored by installing or retrofitting infrastructure with sensors, for example, to monitor the effects of seasonal rainfall in the first few feet of the subsurface. Moisture sensors, resistivity meters, in situ pH sensing with specific ranges, and lysimeters are available or are in various stages of development and could be incorporated into "smart structures" that can monitor for corrosivity. A systems management approach will be needed that can track relevant practices across separately managed infrastructure (e.g., helping the manager of a buried steel structure know if deicing salts applied to pavement by a different infrastructure manager is relevant).

OPPORTUNISTIC DATA COLLECTION

Because excavation is costly, infrastructure owners should take advantage of unexpected opportunities to monitor steel, collect subsurface information, and track infrastructure and subsurface changes.

> **Recommendation 6: Private- and public-sector infrastructure owners should capitalize on opportunities to record properties of the subsurface and steel in a standardized way when infrastructure needs to be maintained, decommissioned, or replaced.**

Opportunistic observations, inspections, and data collection can occur when infrastructure is partially or completely decommissioned (e.g., during maintenance) or replaced. Standardized protocols to collect subsurface property and infrastructure corrosion data should be implemented during those opportunities. Data from fortuitous monitoring opportunities should be systematically saved to inform longitudinal research, DSS, and future decisions for that particular infrastructure, and for buried steel infrastructure more generally.

A DATA CLEARINGHOUSE

Researchers; infrastructure designers, owners, and managers; and steel and steel protection manufacturers could benefit from a public-domain data clearinghouse from which standardized data from multiple industries can be queried and combined to better inform empirically based corrosion rate modeling and corrosion prediction capabilities. The ability to mine data and information would inform corrosion-related experimental designs and modeling and analysis used for infrastructure related design and decision making.

> **Recommendation 7: Industry groups, public-sector agencies with responsibilities related to buried steel infrastructure, and research organizations should coordinate to establish a public-domain data clearinghouse organized around consistent data-recording standards and a common lexicon for secure sharing of data related to the corrosion of buried steel including data on soil environment, corrosion potential and rates, and corrosion monitoring data.**

The data collected as a result of the research and monitoring activities described above need to be made findable, accessible, interoperable, and reusable (i.e., stored as FAIR data) on a public-domain cyber infrastructure platform. A searchable repository of observations and measurements describing the effects of different properties and characteristics of the subsurface on the durability, performance, and corrosion rates of buried steel in a variety of applications is needed on a platform that will provide researchers with data to accelerate fundamental research. This, in turn, will advance the state of the art and of practice in industry. Given proprietary, security, and vulnerability concerns, data could be identified uniquely and without relation to specific location or infrastructure, which may encourage industry to include its data for the benefit of all technical communities. Ultimately, as more data are deposited, advanced data analytical techniques including artificial intelligence and machine learning may be used to mine and analyze the data and enable a more holistic understanding of the environmental contributors to corrosivity and corrosion rates. With an increased availability of standardized, multidisciplinary, and high-quality data collected from well-documented sites, engineering practitioners could investigate and better understand the contributions of combined subsurface properties to corrosivity and corrosion rates in a given type of environment so that future site characterization investigations can be designed more effectively, infrastructure design and management are more efficient, and monitoring programs can target environments and conditions shown to be problematic for certain types of infrastructure. Developing a data clearinghouse is a long-term investment. Experts from appropriate engineering and scientific disciplines, as well as the relevant data scientists, software engineers, information technology specialists, and data curators, should be convened to identify the mission and goals of a clearinghouse, the type of data resources to be made available, the characteristics of the typical data contributors and users, the potential value of the data now and in the future, the infrastructure and personnel necessary to manage the clearinghouse in the short and long terms, the major cost drivers, and the costs of curating and managing data in the short and long terms.

CONCLUDING THOUGHTS

No expert from any sector should be complacent about their assumptions associated with the corrosion of buried steel. Given the complexities of the subsurface environment and the numerous factors that contribute to corrosivity and corrosion rates, improved multidisciplinary understanding of corrosion and corrosivity will yield better decisions related to site characterization, corrosion prediction, steel design and protection, and installation than decisions based on the routine application of higher factors of safety. Better performing infrastructure will result. Industry groups can work through their memberships and with each other to develop research needs statements and calls for proposals based on the above recommendations. Entering into partnerships with each other, the public sector, and academe, a common lexicon can be developed, research goals and priorities developed, research conducted, and data appropriately managed and shared to inform new tools and resources. These recommendations are visionary, and it will be necessary for agencies and organizations to collaborate and coordinate efforts to fund and implement them. However, their implementation is expected to lead to model validation and advances in industry-specific practices. These changes will reduce the costs to safety and the environment and for operation and preservation of buried steel infrastructure.

1

Introduction

Steel is a ubiquitous and important component of the nation's infrastructure, especially infrastructure that is partially or fully constructed beneath Earth's surface (i.e., the subsurface). Steel may be a primary construction material (e.g., as in fuel or water pipelines) or integrated into structures for support (e.g., as in structural foundations, retaining structures, and for roof control in tunneling and mining applications). Corrosion of steel often contributes importantly to infrastructure failure, which can result in loss of life and in adverse and prolonged effects on the environment, public health, safety, and the economy. Repairs associated with corroded steel can be costly (see Box 1.1). Problems associated with corrosion of steel are not new: protection against steel corrosion was the subject of one of the very first committees convened by the National Academy of Sciences (NAS) in 1863 — the year the NAS was chartered. That committee considered methods to protect the bottoms of iron ships from corrosion (True, 1913).

Corrosion can be thought of as degradation caused by environmental conditions. Corrosion mechanisms are well understood in theory, and corrosion protection practices are often employed and effective. However, the subsurface environment (e.g., unconsolidated sediments — called "soils" by geotechnical and structural engineers) in which the steel is placed is spatially and temporally complex and heterogeneous. As such, it is impossible to predict with certainty where, when, and at what rates corrosion will occur, and therefore which corrosion protection measures might be most effective. When unexpected corrosion occurs, it is often impossible to determine whether the corrosion was the result of inaccurate site characterization and prediction of environmental corrosivity, poor choice of steel design or protection, material defects, poor construction or quality control practices, insufficient monitoring or maintenance, changes in subsurface conditions, or some combination of these factors. This report summarizes the deliberations of a National Academies of Sciences, Engineering, and Medicine (the National Academies) ad hoc committee of interdisciplinary experts convened to summarize mechanisms for corrosion of steel within earth materials. The committee was asked to assess current practices for characterizing the subsurface environment for corrosivity and to assess the use, efficacy, and uncertainties of methods used to predict, identify, and monitor corrosion of steel in earth materials at existing and new infrastructure (see Box 1.2). Several sponsors representing different interests, industry sectors, and types of infrastructure came together to support this activity including the American Society of Civil Engineers, Association for Mechanically Stabilized Earth, Federal Highway Administration, International Association of Foundation Drilling, National Science Foundation, U.S. Army Corps of Engineers, and the presidents of the National Academies. The breadth of interests held by the sponsors indicates the importance of the subject matter to the nation.

THE COMMITTEE'S CHARGE AND INTERPRETATION

The National Academies assembled an ad hoc committee of 11 volunteers (see Appendix A for the committee member biographies) to solicit input from the technical community; to examine critically the state of practice and of art in field, laboratory, and modeling methods for characterizing corrosion of steel buried in earth materials; and to identify sources of uncertainty in those practices. Committee members had expertise in civil, corrosion, geoenvironmental, geological, geotechnical, materials, and structural engineering, as well as in geophysical methods, metallurgy, microbially influenced corrosion, and risk assessment. To the committee's knowledge, this study represents the first attempt to identify comprehensively the uncertainties associated with characterizing, modeling, and monitoring corrosivity and corrosion of buried steel across industries by a multidisciplinary group that includes not only experts in corrosion, corrosion engineering, and metallurgy but also experts in civil, geotechnical, and structural engineering.

The committee interprets its Statement of Task (see Box 1.2) as dictating an objective assessment of practices associated with identifying and mitigating corrosion of steel in contact with earth materials throughout the life

---

BOX 1.1
Costs of Corrosion

The annual costs associated with metallic corrosion in the United States have been estimated to be 3–4 percent of the U.S. gross domestic product (GDP) and estimated to be $2.5 trillion of the global GDP (Koch, 2017). Although data are not available to discern the contributions of individual metals to these larger estimates, the contribution of buried steel is substantial.

Local and regional costs of replacing corroded steel can become exorbitant. For example, the Los Angeles Department of Water and Power developed a $1.34 billion plan to replace 435 miles of deteriorating pipe over 10 years (LADWP, 2016), averaging approximately $30 million per mile of pipe. Given the complexity and heterogeneity of the subsurface environment, no methods exist to adequately identify where the worst deterioration is located and where the most resources are needed.

BOX 1.2
Statement of Task

An ad hoc committee of the National Academies of Sciences, Engineering, and Medicine will conduct a study that will solicit input from the geotechnical and civil engineering and materials science technical communities to critically examine the state of knowledge and technical issues regarding the corrosion of steel used in earth applications (e.g., for ground stabilization, pipelines, and infrastructure foundations) and subsurface environments (e.g., unconsolidated earth or rock in different geologic settings). The study committee will

- summarize the primary mechanisms for corrosion of bare or coated steel buried within earth materials under different geologic conditions;
- assess the current state of practice for characterizing native and constructed earth and the subsurface environment for properties that contribute to or influence corrosion and corrosion rates; and
- assess the use, efficacy, and uncertainties associated with methods for predicting, identifying, and monitoring corrosion of steel in earth materials for new and at in-service facilities.

The study will include a workshop on field, laboratory, and modeling methods for characterizing corrosion of steel buried in earth materials and new developments in the prediction and monitoring of corrosion of steel in earth applications and environments. The final report will identify gaps in knowledge and the short- and long-term research needed to improve the long-term performance of steel in Earth applications.

cycle of infrastructure. The practices described are not necessarily state of the art. Materials may be "buried" (e.g., covered with earth material) or "placed" (e.g., drilled, hammered, or otherwise installed) in soils or rock. The relevant infrastructure serves a variety of purposes within different sectors, and, as such, is designed, constructed, and operated by various experts and governed by standards created by bodies serving different industries. Similarly, the extent and consequences of corrosion can vary greatly in different applications. In some civil applications (e.g., the use of steel to support a small embankment), the placement of steel components may be quite localized (e.g., a 10-meter span). In such cases, site characterization and monitoring may be relatively simple, and protection measures against corrosion may be designed specific to the site. Failure of a single component might represent minimal risk to life and safety, and may represent only short-term disruption of infrastructure operation. In other applications—for example, a 100 kilometers-long steel pipeline transporting combustible fuels—the infrastructure may be buried in numerous kinds of earth materials under different environmental conditions. Simple logistics prevent detailed characterization along the length of the pipeline, and failure at any location on the pipeline could result in, for example, loss of life, destruction of property, damage to the environment, and disruption of energy transport serving large populations. Given the array of circumstances in which steel comes in contact with earth materials, an assessment by the committee of all practices for all applications was not a reasonable undertaking.

Early in its deliberations, the committee recognized little common language surrounding key issues related to corrosion, corrosion rates, corrosivity, and corrosion protection among those from different sectors and expertise. Important terms are understood and defined differently across sectors and by professions within sectors, practices and guidance vary greatly between sectors, and knowledge does not often transfer easily across sectors. As a result, the committee found it necessary to identify and define the relevant types of steel, the different subsurface environments and conditions that affect corrosivity, the range of buried-steel applications and protection measures, and even the fundamental types of steel corrosion so that committee members had a common basis of understanding. Much of the early text of this report is primer material developed by the committee so that its own members could deliberate effectively and meaningfully. Readers of this report will need familiarity with this material to understand and apply the report recommendations and should not assume that their familiarity with specific vocabulary and concepts is consistent with terms and concepts as defined in the report.

To focus the broad task assigned to the committee, the report does not cover "best practices," steel in concrete (except reference to cementitious grouts), steel in marine environments, or interior pipeline corrosion. Steels protected with grouts are considered because grouts form a low-volume matrix that is integrated into the soil in contact with the buried steel. Stainless steels are mentioned in this report but not considered extensively.

COMMON TYPES OF STEEL USED IN BURIED APPLICATIONS

Steel is a metal composed of iron mixed with various metallic (e.g., manganese) and nonmetallic (e.g., carbon) elements in low concentrations to create alloys fit for different purposes. With careful control of composition and processing, steels can possess a wide range of properties such as strength, ductility, and corrosion resistance. The choice of steel for a particular application is based on the properties required, cost, and availability. The wide variety of steels available can be classified differently, including based on composition, properties (e.g., strength and ductility), and application. Steels of interest in this report are those placed commonly in the subsurface. These include plain carbon steel (including plain low-carbon or mild steel, plain medium-carbon steel, and plain high-carbon steel), high-strength low-alloy steel, and cast iron/ductile iron (see Table 1.1 for descriptions). These classifications are based on the composition, including the content of carbon and other alloying elements. Steels may also be classified according to yield strengths before nonrecoverable deformation occurs. Steels may be assigned grade numbers to reflect the minimum yield strength in units of kilopounds per square inch (ksi), regardless of composition.[1]

The inclusion of carbon in steel strengthens it but increases its brittleness and decreases its weldability. This makes optimization of the carbon content dependent on the application. Most structural applications use plain low-

---

[1] Common steel grades used currently in subsurface applications are Grades 50, 65, and 80. Grade 36 steel was used historically for some buried steel components.

TABLE 1.1 General Classification of Steels

Steel Type	Common Name	Alloy Composition	Underground Use	Comments
Plain carbon steel[a]	Plain low-carbon steel or mild steel	<0.3 wt % carbon	Common; structural, pipelines	Often has a protective coating in geo-civil industries.
	Plain medium-carbon steel	0.3–0.6 wt % carbon	Pipelines, structural bolts, bolted connections, splices of structural components	Pipelines are typically coated with bonded dielectric coatings and protected by cathodic protection.
	Plain high-carbon steel	0.6–1 wt % carbon	High-strength steel bars used in anchorages	High strength but low ductility, prone to brittle-type failure.
Cast iron	Gray iron	2.5–4.0 wt % carbon, 1.0–3.0 wt % silicon	Pipelines installed in the United States circa 1810–1970s	Corrosion of iron occurs preferentially to carbon leaving a graphite skeleton. It may come with black asphaltic coating.
	Ductile iron	3.0–4.0 wt % carbon, 1.8–2.8 wt % silicon	Pipelines introduced in the United States circa 1955 as an improvement to gray iron	Usually comes with black asphaltic coating and wrapped in polyethylene encasement (ANSI/AWWA C105/A21. 2018).
Alloy steel	High-strength low-alloy steel	Low carbon and small amounts of cobalt, nickel, chromium, molybdenum, vanadium, nitrogen, etc.	Pipelines	More corrosion resistant in some environments (Fletcher, 2005; Shreir et al., 1994) but susceptible to hydrogen embrittlement.
	Stainless steel	>11 wt % chromium and often other elements	Rare	Very corrosion resistant in some environments.

[a] Often contain small amounts of manganese, phosphorus.
SOURCES: Stefanescu (1990); Washko and Aggen (1990).

carbon steel (also known as mild steel). This type of steel contains less than 0.3 percent carbon. Higher-strength steel such as high-strength low-alloy steels may sometimes be used. Cast iron often has a carbon content above 2.5 percent and is commonly used for water distribution pipelines. Ductile iron is an innovative type of graphite-rich cast iron, with graphite incorporated into the metals in nodules rather than flakes, as is the case in cast iron. Carbon included in this manner provides improved impact and fatigue resistance.

"Alloy steels" are those that contain major percentages of alloying elements in addition to carbon. Stainless steel, for example, results when chromium is included in percentages higher than 11 percent. There are numerous types of stainless steel distinguished by concentrations of carbon, chromium, nickel, molybdenum, manganese, titanium, and other elements. Stainless steels are sometimes used in buried-steel applications, but initial costs are prohibitive, and there is little guidance on the selection of stainless steel type given for the environment and application. None of the common stainless steels are completely resistant to corrosion, and failures have occurred when used improperly. Some governing bodies do not recommend the use of stainless steels (e.g., AASHTO, 2020).

Other alloy steels are used in an array of buried infrastructure applications, from general structural to welded pipe, welded wire, high-temperature applications, welded and bolted connections on buildings and bridges, and soil reinforcements (see Table 1.1). Many civil engineering application designs generally specify alloys that have performed well previously and are offered "off the shelf" by manufacturers. In contrast, pipeline industries often use steels developed for specific applications. Except high-strength steels, which are susceptible to hydrogen embrittlement, the committee could find no evidence in the literature of significant differences in corrosion rates of the different types of plain carbon steels in buried applications (see Chapter 4). Alloy composition is considered of secondary importance to corrosion rates when compared to variable corrosivity found in natural and engineered subsurface environments.

INFORMATION GATHERING AND REPORT ORGANIZATION

The study committee was convened in 2020, and all committee information gathering and deliberations were conducted remotely as a result of the ongoing COVID-19 pandemic. Although the committee only met remotely, members were able to draw information from multiple resources including from many experts with a variety of expertise. The committee held two information-gathering sessions, including a 2-day virtual workshop. Agendas for those meetings can be found in Appendix B. In addition to information gathered through talks and panel discussions during those meetings, the committee relied on its collective and extensive expertise, held one-on-one discussions with experts on a variety of topics, and consulted the published and unpublished literature, instrumentation manuals, and the standards and regulations established for multiple industries.

As discussed earlier, corrosion scientists, geotechnical engineers, and structural and civil engineers have developed their knowledge of corrosion of buried steel and corrosion protection almost independently, and their vocabularies have evolved in response to practices in each of the different fields. Because the committee had to develop a common vocabulary to deliberate its charge and prepare this report, it was necessary for the committee to go back to the very basics. This report reflects that need and provides definitions of specific terms that may not be consistently defined in different sectors.

Chapter 2 of this report provides a description of the fundamentals of corrosion and the committee's observations about the general approaches taken by different industries to address corrosion of buried steel. The text describes some fundamental vocabulary associated with corrosion, the understanding of which was critical to committee deliberations. The text also describes different buried-steel applications, and the sources of data that inform industry knowledge and decision making related to the design and protection of buried steel infrastructure. Chapter 3 describes the complexities of the natural and engineered subsurface environments that contribute to the corrosivity of the environment and the difficulties in characterization. Chapter 4 describes different mechanisms for the corrosion of buried steel and how that corrosion is manifested on the steel surface, while Chapter 5 describes different industry methods to protect against that corrosion. Chapter 6 provides a description of different laboratory- and field-based methods available to characterize the subsurface for corrosive conditions, and Chapter 7 provides a description of standard infrastructure monitoring techniques and practices. Chapter 8 provides descriptions of different modeling techniques applied in different industries. Chapter 9 provides an assessment and conclusions regarding current and emerging practices, and provides recommendations regarding how practice could be improved and research that might be undertaken to increase basic understanding of corrosion of buried steel that leads to better infrastructure-related decision making.

2

Fundamentals of Steel Corrosion,
Industry Applications and Approaches,
and Sources of Corrosion Data

Corrosion of steel is not complicated and is theoretically predictable at the electrochemical level if the steel and the environment in which the steel is placed are understood. However, at any scale in which there is unpredictable or unmeasurable variation in properties that contribute to corrosion—whether in properties of the steel itself (e.g., imperfections on the steel surface) or of the environment in which the steel is placed—the prediction of steel corrosion becomes complicated. Unpredictable spatial and temporal variations can result in severe corrosion and infrastructure failure and are found in the air that surrounds aboveground and aqueous or marine steel installations. However, the spatial and temporal variations of materials found under Earth's surface in which steel is buried are even more complex and difficult to understand, and the steel more difficult to protect.

Before presenting information about the complexities of corrosion and corrosion prediction in later chapters, this chapter presents some fundamentals regarding steel corrosion at the electrochemical level, describes different types of buried-steel applications, discusses general industry approaches to protection of steel from corrosion, and highlights sources of data regarding corrosion of buried steel. More details about the subsurface environment and its variations are found in Chapter 3. Descriptions of different mechanisms of corrosion are presented in Chapter 4.

FUNDAMENTALS OF BURIED STEEL CORROSION

Extraction of elemental metals from ores requires the input of large amounts of energy. As such, the elemental metals and metallic alloys are thermodynamically unstable and tend to revert to stable oxides similar in form to their ores found in nature. Therefore, from a thermodynamic perspective, it is logical to expect that steel—whether buried or exposed to the atmosphere or some other environment—will corrode. The corrosion of steel buried in soil, rock, or grout occurs predominantly through an electrochemical process involving three key components: an anodic site of the steel (where the steel is electrochemically active—oxidation occurs at this site and electrons are lost), a cathodic site of the steel (a less active region—reduction occurs at this site and electrons are gained), and an electrolyte in contact with the two sites, which completes the circuit by conducting ionic current between the anodic and cathodic sites. The soil, rock, or grout in which the steel is exposed can be the electrolyte if it is sufficiently wet and conductive. The scales at which corrosion is facilitated can range from less than a nanometer to more than a meter, depending on material and environment. In other words, in essentially all instances, if steel is buried in these materials, the condition exists for corrosion to occur. The focus in this report hereafter will be on soil as the electrolyte.

At the anodic sites, iron atoms [Fe^0 or Fe] are stripped of electrons to form iron ions, Fe^{2+}:

$$Fe \rightarrow Fe^{2+} + 2e^-$$

Equation 2.1

This corrosion reaction causes the loss of material, and the reaction is an oxidation (or anodic) reaction because electrons are generated. It is a half-cell reaction that must be accompanied by one or more reduction (or cathodic) half-cell reactions that consume the electrons to uphold the requirement of charge conservation.[1] For corrosion in wet environments, such as in moist soils, the typical cathodic reactions that occur at the cathode are oxygen reduction and hydrogen evolution reactions. If dissolved molecular oxygen [O_2] is present in the soil electrolyte (oxic conditions), it can be reduced to form hydroxyl ions [OH^-]:

$$O_2 + 2H_2O + 4e^- \rightarrow 4OH$$

Equation 2.2

The presence of oxygen will also further oxidize the dissolved Fe^{2+} ions to create various types of hydroxides or oxides, such as hematite, Fe_2O_3, that form the familiar red rust corrosion product so strongly associated with corroded steel:

$$4Fe^{2+} + O_2 + 4H_2O \rightarrow 2Fe_2O_3 + 8H^+$$

Equation 2.3

If there is no dissolved oxygen in the soil electrolyte (anoxic conditions), the electrons generated by the anodic dissolution reaction can be consumed by the hydrogen evolution reaction through the reduction of water to form hydrogen gas [H_2]:

$$2H_2O + 2e^- \rightarrow H_2 + 2OH^-$$

Equation 2.4

Another form of the hydrogen evolution reaction involving the reduction of H^+ ions might occur in acidic soils where these ions are in abundance:

$$2H^+ + 2e^- \rightarrow H_2$$

Equation 2.5

Electrons flow through the steel between the anodic and cathodic sites on the steel, and ions released as a result of the reactions are transported through the electrolyte to complete the electrochemical circuit. The driving force for corrosion is the corrosion potential—or the measured voltage between the steel surface and the electrolyte. Box 2.1 provides a description of corrosion potential. More details of steel corrosion are provided in Chapter 4.

BURIED-STEEL APPLICATIONS

Soil and rock serve two primary but distinct functions in infrastructure construction: first, they can be used as building materials to construct geotechnical assets (e.g., embankments, dams, slopes, and some types of retaining walls); and second, they can be used as foundations to support load and to limit settlement of surficial or buried structures (e.g., building foundations or buried tanks). In the first case, soil and rock are used as construction materials, while in the second case, soil and rock are the foundation materials on or in which construction is performed. The role of buried steel will be different depending on the type of construction application. For example, when soil is used as a construction material in an earthen retaining wall, steel strips may be embedded in the soil to carry tensile load within the structure of the wall (see Figure 2.1A). This is because the tensile strength of soils is effectively zero even though soils can carry compressive and shear loading. When used as foundation materials,

---

[1] A half-cell generally refers to one part of two half-cell readings that comprise an open circuit of corrosion potential. Different technical communities might not consider "half cell" as an acceptable equivalence to open-circuit potential, but the term is used here because it is more commonly used in the geotechnical community.

BOX 2.1
Corrosion Potential Versus Potential for Corrosion

In the geotechnical field, the term "corrosion potential" is used often to describe the potential or possibility of corrosion that compromises structural integrity. Indeed, the likelihood of steel corrosion to occur will vary depending primarily on the details and the corrosivity of the environment to which it is exposed. However, in the field of corrosion engineering, the term "corrosion potential" has a different and specific scientific meaning. It is an electrochemical potential and refers to a measurable voltage associated with the separation of charge at the interface of the metal surface and the electrolyte. It also is variably referred to as open-circuit potential, free potential, or half-cell potential.

Water molecules and the dissolved positive and negative ions are randomly oriented and distributed when distant from metal placed in a corrosive environment such as soil. Closer to the metal, the water and ions spontaneously rearrange, creating a separation of charge—or a potential drop—across this near-surface region (see Figure 2.1.1). The extent to which this happens can be measured with a voltmeter and a second electrode called a reference electrode. This measured voltage is called the corrosion potential (denoted as E_{corr} and measured in volts relative to the specific reference electrode). E_{corr} is an important parameter because, in simple terms, it is the driving force for corrosion. In this report, the term "corrosion potential" is reserved for the description of the potential drop at a metal surface, E_{corr}. This report uses the term corrosivity when discussing the environment. This was one of many terms with different usage among geotechnical and corrosion engineers that the committee found to complicate effective communication about the issues of corrosion of buried steel.

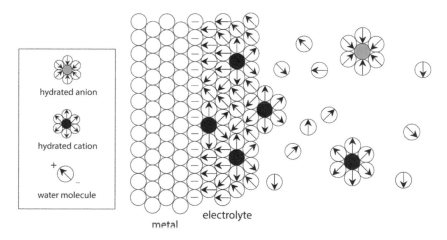

FIGURE 2.1.1 Schematic representation of the separation of charge at the interface of a corroding metal and an aqueous electrolyte. The anions and cations are charged ionic species. The arrow in the symbol for the water molecule represents a charge dipole caused by the structure of the water molecule, H_2O. The spatial separation of the ions and the alignment of water dipoles results in a potential drop from the metal surface into the solution, which is associated with the corrosion potential.

FIGURE 2.1 Examples of buried steel infrastructure. **(A)** Steel reinforcement strips used in mechanically stabilized earth walls. **(B)** Steel H piles for construction of the Henderson Bridge. **(C)** A sheet pile barrier being installed to keep rising Mississippi River water away from oil-contaminated wetlands at the Galena Train Derailment in Illinois.
SOURCES: Photos from the **(A)** Reinforced Earth Company; **(B)** City of East Providence, Rhode Island (2021); and **(C)** EPA (2015).

soil and rock provide strength to prevent structure failure and stiffness to resist excessive structure settlement. For example, when a deep foundation is constructed using a steel H pile (so-called because the cross section is shaped like an "H"; see Figure 2.1B), the soil exerts resisting friction on the side of the buried pile, preventing it from settling more than specified by the design criteria. In many cases, infrastructure is designed to transmit loads exerted by the structure through steel components to the surrounding soil, which requires interaction through intimate contact between the steel and soil to prevent failure and limit settlement.

Manufactured steel components are designed with specific materials and defined shapes for desired properties. However, during exposure to the environment during service life, those properties can diminish as a result of corrosion. Most metals, including iron used in engineering applications, are refined from naturally occurring ores (i.e., oxides, carbonates, hydroxides, and silicates). The interaction between the subsurface environment (with its various solid, liquid, or gas phases) and the steel will determine the occurrence, extent, and rate of corrosion, but it is important to note that the design of the buried steel contributes to how much surface area contact there is between the subsurface and the steel and therefore how much of the steel is exposed to corrosive environments. Circular steel bars, for example, have a low ratio of surface area to volume while thin sheet piles (see Figure 2.1C) have a high ratio of surface area to volume. In contrast, the surface area contact between the subsurface and pipelines is dominantly dependent on the quality of installation and the technique used. Installation of pipelines can occur in an excavated trench where pipe will be placed in contact with backfill. This backfill can have a highly variable degree of quality control depending on the uniformity of soil compaction around the pipe and the presence or absence of air gaps on the surface of the pipe larger than the pore spaces of the soil. Alternatively, pipe can be placed through horizontal directional drilling, where it will have less uniform contact with the native soil and possibly non-native drilling mud. Table 2.1 outlines common types of infrastructure that include buried steel components.

GENERAL INDUSTRY APPROACHES TO PROTECTION AGAINST STEEL CORROSION

The committee identified two general approaches to corrosion protection applied to buried steel. These tend to drive methods of site characterization and infrastructure design but may also contribute to what the committee identifies as "silos" in practice and corrosion-related research. One approach compensates for corrosion by designing for the expected corrosion-related loss of metal over time. It is applied, for example, to structural foundations, earth retaining structures, dams, and tunnels. The committee calls this the "corrosion allowance approach" and calls those industries in which this approach is applied the "geo-civil industries." The geo-civil industries attempt to predict corrosion rates by characterizing the corrosivity of the earth materials in which the steel is placed. The steel components are then designed with sufficient volumes of steel to compensate for expected steel loss due to corrosion over the design lifetime of the component, the application of protective coatings (e.g., galvanized zinc-coated steel reinforcing strips used in mechanically stabilized earth [MSE] walls), or both. These designs include redundancies in the system so that failure of one component does not result in catastrophic failure of the system.

The other approach to corrosion protection focuses on the prevention or reduction of the rate of corrosion through such means as coatings and cathodic protection. The oil and gas pipeline industries typically apply these approaches and tend to consider the life cycle and effectiveness of the prevention measures. Those industries that apply what this report calls the "corrosion prevention approach" are less interested in the lateral and vertical variations in corrosivity of the subsurface environment and more interested in controlling the rate of corrosion.

Not all industries or applications fit cleanly into one of these approaches, and there are exceptions within the applications described above. Some designs may use hybridized approaches, or an industry may use one or the other approach in a given application. However, this delineation of approaches is a useful way to generalize and categorize industry practices. In all cases and industries, the corrosion of buried steel is caused by the same mechanisms, and the subsurface characterization techniques used by the different industries might be combined to inform both modeling of corrosion rates and design of protective measures.

MICROORGANISMS

The presence of microorganisms and microbial activity can be anticipated on most buried steel infrastructure. The microbial community controls the chemistry at the soil–steel interface and can affect corrosion. All subsurface environments that contain water, nutrients, and energy sources in addition to electron donors and acceptors are biologically active. Essential nutrients include assimilable forms of carbon, nitrogen, phosphorus, sulfur, and trace elements. Respiration provides cellular energy through a series of coupled oxidation and reduction reactions involving the transfer of electrons from donors to acceptors. Electron donors can include organic carbon compounds (e.g., acetate, lactate, glucose), H_2, or iron (Fe^0). In aerobic respiration, the electron acceptor is oxygen (O_2). In anaerobic respiration, electron acceptors are molecules other than oxygen (e.g., CO_2) or ionic species (e.g., sulfate, nitrate, Mn^{+4}, and Fe^{+3}). With a given electron donor (e.g., glucose) aerobic respiration yields more energy than anaerobic metabolism.

Subsurface microorganisms can form biofilms directly on buried steel surfaces. Biofilms contain living and dead cells enmeshed in a matrix of extracellular polymeric materials and can range in thickness from a few to several hundred micrometers. A property of undisturbed biofilms in oxygenated environments is stratification of electron acceptors with depth (Nealson and Stahl, 1997). A typical progression of stratification begins with oxygen depletion and ends with sulfate reduction or carbon dioxide reduction. The reduction of sulfate produces sulfide and the reduction of carbon dioxide (i.e., methanogenesis) produces methane (Nealson and Stahl, 1997). The thickness of biofilm required to produce an anoxic environment depends on the respiration rates of organisms in the biofilm and the availability of oxygen in the environment. Aerobic respiration rates can be faster than the rate of oxygen diffusion (Yu and Bishop, 2001). In the presence of a biofilm, steel may be subject to anaerobic corrosion processes when the bulk environment is oxic. Microorganisms in biofilms do not behave as individuals. Instead, microorganisms engage in cooperative activities by responding to diffusible molecules and produce stratified communities based on electron acceptor availability.

Soils contain an enormous diversity of microbial communities that are metabolically versatile and adaptable. Soil microorganisms play crucial and often complementary roles in nutrient cycling, processes that can include biodegradation of complex carbon compounds, nitrogen fixation, and phosphorus solubilization (Barros, 2021). As an example of extreme microbial adaptation, Ortiz et al. (2021) identified microorganisms in Antarctic desert soils that were capable of harvesting solar energy, oxidizing inorganic substrates, and adopting symbiotic lifestyles.

Bacteria are the microorganisms that have received the most attention in the characterization of subsurface environments for corrosivity, but representatives from the three branches of life (bacteria, archaea, and eucaryotes [fungi]) have been identified in soils and rocks (Banciu, 2013; Bintrim et al., 1997). Microbial growth may occur on the surface of rocks or in crevices or fissures centimeters deep within the rock (endolithic) (Gorbushina, 2007). While microbial population sizes, distributions, and growth rates can be determined by the availability of water (Stevenson et al., 2015), endoliths have developed several survival strategies even in non-optimum conditions. For example, some microorganisms can survive long periods of starvation and desiccation, and their spores can remain viable for hundreds of years. In summary, microbial activity can be present in a wide range of geologic settings, and microbes are adept at adapting to different environments.

TABLE 2.1 Common Types of Infrastructure That Use Buried Steel

Type of Infrastructure	Definition and Function	Construction and Design
Deep foundations	**What are they?** Deep foundations are long, slender infrastructure components (depth > diameter) designed to transfer loads from infrastructure to soil or rock. Used when loads cannot be sufficiently supported, when the foundation must penetrate water, or when excessive settlement would result from using shallow foundations. **How common?** Most bridge foundations, taller structures (>5 stories), and offshore structures incorporate deep foundations. **Corrosion implications?** When deep foundations corrode, they lose ability to carry load; structures tilt or collapse.	**Piles:** Driven or vibrated into the ground. All piles are designed for intimate contact with soil and may bear on or into rock. Steel pile soil–steel interfacial contact area will be a function of geometry and influences corrosion. Corrosivity is addressed by increasing steel cross sections (i.e., thicker steel) or through addition of protective coating. **Drilled shafts:** Constructed by drilling a hole larger than 30.5 cm (12 inches) in diameter, inserting steel, and pouring in concrete. Steel is required for structural reinforcement and may corrode if the concrete cracks. **Micropiles:** Constructed by drilling a hole with a diameter less than 30.5 cm (12 inches), inserting steel, and pumping in grout. Steel casing may also be inserted in the upper portion of the hole and grout pumped through the center. Steel casing will be in direct contact with the soil, making corrosivity a significant design consideration that can be addressed with thicker steel casing cross sections or protective coating of steel. The reinforcement may corrode if the grout cracks.

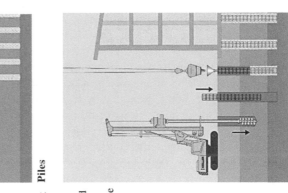

Piles

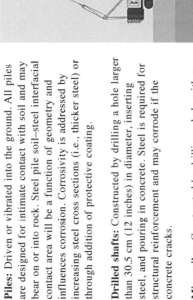

Drilled Shaft

continued

TABLE 2.1 Continued

Type of Infrastructure	Definition and Function	Construction and Design
Earth retaining structures	**What are they?** Designed to resist lateral load and to retain soil and groundwater. Can be rigid or flexible and are generally constructed of concrete or steel, with soil as an integral part of the design. Drainage is a critical stability consideration in most retaining structure designs—most designs include drainage of the retained soil. **How common?** Areas with grade changes that must be controlled for infrastructure (e.g., highway embankments, building foundation slabs). Also used for support of excavations and along waterfronts to harden the shoreline or contain reclaimed land. **Corrosion implications?** When steel components in earth retaining structures corrode, they lose the ability to resist lateral loads and retain soil, and structures will deform, which can lead to collapse of the structure and supported infrastructure.	**Mechanically stabilized earth (MSE) walls:** Designed to use friction between the soil and long thin steel or polymeric strips that are embedded horizontally behind the wall face to prevent deformation. The strips are connected to the wall face and are designed to resist lateral movement of the wall. Corrosivity is a significant design consideration for the steel strips—often low-corrosivity soils are selected, and thicker steel cross sections are used with protective coatings (e.g., galvanized steels). **Sheet piles:** Driven, pushed, or vibrated thin steel sections frequently used in soft soils. Sheet piles are in intimate contact with the soil being retained. Corrosivity is addressed with thicker cross section (i.e., thicker steel). 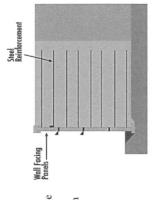 **MSE Wall** 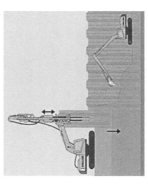 **Sheet Pile**

Steel Reinforcement

Wall Facing Panels

MSE Wall

Sheet Pile

continued

TABLE 2.1 Continued

Type of Infrastructure	Definition and Function	Construction and Design
Ground anchors (also known as tiebacks)	**What are they?** High-strength steel bars or cables used to transfer tensile loads to stabilize ground. **How common?** Used when additional support is needed to resist undesired movement of a retaining wall face; to uplift from tall slender structures such as towers; or to add stability for dams, tie downs, retaining walls, or foundation systems. Often used in retrofit applications. **Corrosion implications?** When the steel cable corrodes, it loses the ability to maintain the tensile load, which will allow undesired movement. Excessive movement can lead to failure.	**Ground anchors/tiebacks:** Constructed by drilling a hole through the unstable portion of the soil or rock mass and into the underlying stable zone, inserting a steel cable or bar, and grouting in place. A steel sleeve can also be used around the steel above the grouted stable zone so the steel can be tensioned and connected to the structure. Corrosion protection is sometimes referred to as double corrosion protection since the steel is encapsulated in grout and also encased in a sleeve. 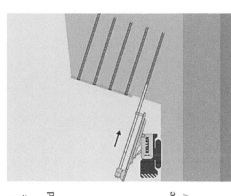 **Ground Anchors**
Soil nails	**What are they?** Closely spaced tension-carrying steel bars grouted into soil. **How common?** Used as a retrofit to stabilize slopes/landslides or to add lateral support to tall excavations and retaining walls (>5 stories). **Corrosion implications?** When the steel corrodes, it loses the ability to maintain the tensile load, which will allow undesired movement that may cause failure. Excessive movement in a retaining wall or excavation can lead to failure.	**Solid-bar soil nails:** Constructed by drilling a hole, inserting a steel bar, and pumping in grout along the length of the bar. After installation, soil nails are covered with pumpable, sprayed concrete and capped with bearing plates. The grout mitigates corrosion via encapsulation. Steel may be coated with zinc (galvanization) for additional corrosion protection. **Hollow-bar soil nails:** Installed in a single-step process where the bar is used to advance the hole, and grout is injected through the hollow center of the bar. Additional corrosion mitigation from galvanization (zinc coating) is not feasible due to the way the bars are installed. Corrosion is addressed by increasing the bar cross section. **Soil Nails**

continued

continued

TABLE 2.1 Continued

Type of Infrastructure	Definition and Function	Construction and Design
Rock bolts/shear pins	**What are they?** High-strength steel bars or cables that use tensile forces to anchor unstable rock masses. **How common?** Used when a slope or cut slope in a rock mass is unstable. Used as part of tunnel support systems and as roof supports in underground mines. **Corrosion implications?** When the steel corrodes in a section under tension, it cannot stabilize the mass and the rock face or slope will move (or fail).	**Rock bolts:** Tensioned bolts for use in an unstable rock mass. Constructed by drilling a hole, inserting the steel, and pumping grout along the length of the steel. Grout is used for stability in the stable rock and to control corrosion due to water seepage. **Shear pins:** Untensioned pins for use in an unstable slope. Constructed by drilling a hole into the rock, and inserting and grouting a steel rod. The steel rods mitigate slip by resisting shear through their cross section but do not undergo tension that would otherwise add to a frictional resistance due to an increase in normal stress across the shear plane (i.e., shear pins offer shear resistance but do not otherwise affect the state of stress within the rock). 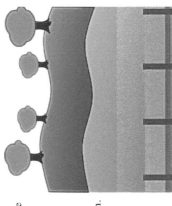 **Rock Bolts**
Pipelines	**What are they?** Long, linear infrastructure networks used to transport gases and liquids great distances (in some cases thousands of kilometers). Installed by trenching or pushing the pipe into the soil from access trenches using hydraulics (i.e., pipe jacking), or by pulling pipe through a drilled hole. **How common?** Most water, oil, and gas are transported through a pipeline. **Corrosion implications?** Characterization and control of the burial environment is limited due to length. If a water pipeline corrodes, there can be a loss of structural support and transported liquid can leak. Water leaks can cause a localized loss of water pressure or remove any surrounding soil or infrastructure such as a road. Corrosion of oil or gas lines can cause loss of structural support or a catastrophic failure, sometimes leading to environmental damage or an explosion.	**Water lines:** Primarily manufactured from cast or ductile iron. Corrosion is addressed by increasing the cross section of the pipe (i.e., steel thickness) based on the soil corrosivity. **Oil or gas lines:** Primarily manufactured from plain low-carbon steel. Coatings and cathodic protection are used across pipeline systems to address corrosion. **Pipeline**

TABLE 2.1 Continued

Type of Infrastructure	Definition and Function	Construction and Design
Underground storage tanks	**What are they?** Cylindrical or rectangular prisms or other shaped steel containers used for storing liquids such as water and fuel. **How common?** Not as common as other types of buried steel. Common application has been at gas stations but is becoming less frequently used compared to inert polymer and plastics. **Corrosion implications?** Corrosion can result and leakage, depending on the type of liquid, can result in environmental contamination including groundwater.	**Underground storage tanks:** Fuel and water storage tanks are made of ductile iron or mild steel. Tanks are generally made by welding steel plates together to form the desired shape of the tank. **Underground Storage Tank**

SOURCE: Diagrams of piles, drilled shafts, MSE walls, sheet piles, ground anchors, soil nails, and rock bolts from Keller Management Services, LLC (2022).

SOURCES OF CORROSION DATA

Long-term in situ testing to understand and measure the corrosion of buried steel is limited, but the most well-known studies were performed from 1922 to 1940 by the National Bureau of Standards (NBS, now National Institute of Standards and Technology) (Logan, 1945; Romanoff, 1957). In these studies, samples of uncoated alloy steel and wrought iron (i.e., low-carbon steel) were buried at 47 locations throughout the United States. The locations were chosen to test corrosion of steel in a range of soil types and climates. To mimic real-world conditions, the long-term experiments performed by NBS were designed to test steel in field conditions with no control or alteration of the soil properties. The steel was allowed to corrode in nonuniform soil conditions. Interpretations of the resulting data indicate that corrosion rates typically attenuated with respect to time, depending on the subsurface conditions.

The NBS studies had several experimental design limitations. Soil environments were not thoroughly characterized, burial depths and exposure times were variable, exposure to moisture was not monitored, rainfall was not quantified, climatic variables were reported as averages, many soil properties were measured off-site, and the statistical design of the study was weak (de Arriba-Rodriguez et al., 2018). Despite experimental limitations, the NBS data still are the basis for steel corrosion rate predictions today. It is reasonable to conclude that the high level of uncertainty associated with the experimental data could lead to either costly overconservative design or unsafe underconservative design.

Since completion of the NBS studies, others have studied the performance of buried steels and archived data from specific applications (e.g., related to MSE walls, piles, ground anchors, rock bolts, and culverts). Some of those studies are described in the list below, and data from them may be used to better understand the reliability of existing corrosion models for estimating service life or the factors that contribute to corrosivity in different applications. Although these studies are not as comprehensive as the NBS efforts, the resulting data supplement the NBS data and are applicable to specific sets of conditions that were not addressed in the NBS study. For example, less than 10 percent of the NBS data were from free-draining granular soils such as those used in constructed earth applications. Even fewer of the data are related to corrosion of galvanized steels. Below are some additional sources of data that supplement the NBS database.

- Corrosion of metallic earth reinforcements in MSE wall backfill is addressed with a 20-year study by Darbin et al. (1986). NBS data (Romanoff, 1957) on the performance of MSE reinforcements were supplemented by industry research (Darbin et al., 1986; Haïun et al., 2007), by research of proprietors of MSE wall systems in Europe (France) beginning in the 1970s, and with results from recent research sponsored by the National Cooperative Highway Research Program (Fishman and Withiam, 2011).
- Corrosion, durability, and service-life modeling for steel piles including steel (mostly), reinforced concrete, and prestressed concrete piles have been described by Schwerdtfeger and Romanoff (1972), Statfull and Seim (1979), Ohsaki (1982), Long (1992), Long et al. (1995), Beavers and Durr (1998), Törnqvist and Lehtonen (1999), Wong and Law (1999), Decker et al. (2008), Sagues et al. (2009), Gu et al. (2015), and Poursaee et al. (2019). (Note that reinforced concrete and prestressed concrete piles are not covered in the present report.)
- Data related to the corrosion of various types of steel piling exposed to underground conditions at 53 different sites are included in NBS Monograph 127 (Schwerdtfeger and Romanoff, 1972). The steel was buried in climate ranging from semitropical to frigid, above and below the water table, and in soils encompassing a broad range of corrosivity.
- Data related to pile performance from seven sites located in New York, Connecticut, Maryland, North Carolina, and Mississippi were collected by Beavers and Durr (1998). In all cases the subsurface was layered, and corrosion observed along the piles was confined to one of the layers. The highest corrosion rates were observed within non-natural fills such as ash, slag, and cinders where the interface between native soils and non-natural fills was at or near the water table. The results reported by Beavers and Durr

demonstrate how site conditions in addition to the electrochemical properties of the earthen materials have a significant effect on corrosivity.

- Data from four cases where severe corrosion of steel piles advanced through industrial fill including slag and cinder ash with groundwater near the base of the fill are reported in Connecticut (Long et al., 1995), in New York (Moody, 1993), in Pennsylvania (Gu et al., 2015), and by the Wisconsin Department of Transportation (WisDOT, 2015). In cases where piles have not been advanced through industrial or engineered material, the observed corrosion is low.

- Corrosion rates and predictions based on the long-term field performance of piles in Utah were evaluated by Decker et al. (2008). Twenty piles were extracted from five sites after service lives of 34–38 years. Measurements included soil index properties, resistivity, pH, cation/anion concentrations, and water table elevation.

- The possibility of correlating water pH, soil resistivity, and soil corrosion potential with the durability and corrosion of culverts was investigated by Boyd et al. (1999). Fifty-one sites with a variety of culvert types were included in the study. Variables including soil resistivity, flow rate, and streambed geometry were measured, and descriptions of the pipe slope, presence and speed of water, streambed type, and other parameters were recorded.

- Possible correlations among bed load, water velocity, corrosion rate, culvert installation design, and site conditions are described by Ault and Ellor (2000). They studied the durability of corrugated (mostly aluminized Type 2; see Chapter 3) metal pipe. Field and laboratory studies included 32 culverts in three states. The effects of a number of different parameters were investigated, including pipe size, slope, and corrugation type; soil temperature, resistivity, pH, and abrasiveness; and water temperature, resistivity, pH, chemistry, and flow rate.

- Studies on the corrosion of ground anchors, rock bolts, and soil nails are reported by the National Cooperative Highway Research Program (Withiam et al., 2002), the Fédération Internationale de la Précontrainte (1986), the New Hampshire Department of Transportation (Fishman, 2005), the New Brunswick (Canada) Department of Supply and Services (Snyder et al., 2007), the Washington State Department of Transportation (Kramer, 1993), the Federal Highway Administration (Cheney, 1988), the Texas Department of Transportation (Briaud et al., 1998; Gong et al., 2006), and the Geotechnical Engineering office of Hong Kong (Shiu and Cheung, 2008), and for sites in Japan (Tayama et al., 1996) and Scandinavia (Baxter, 1996; Lokse, 1992). Other studies were conducted by Sundholm (1987).

- Reported incidents of ground anchorage failures caused by corrosion compiled by the Fédération Internationale de la Précontrainte (1986) were summarized by Littlejohn (1992). Thirty-five case histories comprising 24 failures of permanent installations are included. Based on the results of reported incidents relative to the number of anchors installed during the 31-year time frame covered by the survey, the incidents of prestressed ground anchor failures by corrosion were limited and generally random, with the possible exception of steel-type (grade, manufacturing process, etc.). Corrosion was localized and appeared irrespective of tendon type (i.e., bar or strand). Corrosion generally occurred where the tendon steel intersected a crack in the protective grout. Cracks in the free and fixed anchor zone were formed either as the result of shrinkage strains due to curing, or the result of tensile loading.

- Data on rock bolt performance, mostly from underground mining and tunneling applications in Scandanavia, Australia, Canada, and the United States are summarized by Kendorski (2000). Kendorski compared performances considering the effects from different grouts used to install the anchorages, and for different types of rock bolt and ground anchor systems.

3

Subsurface Environment

The subsurface environment in which steel is buried or placed impacts the rate and extent of corrosion. The subsurface is often a heterogeneous mix of solids, liquids, and gases, with variations in soil and rock composition, moisture content, groundwater flow, and the microbial activities that lead to the chemical, electrochemical, or biological process that affects corrosivity. Variations in those properties occur temporally and spatially. Several parameters have been identified that influence the corrosivity of a subsurface material, including its ability to conduct electricity (i.e., electrical conductivity), the degree to which the soil is saturated with water, pH, dissolved salts, oxidation-reduction potential (also called redox potential for reduction-oxidation)—a measure of the oxidizing power of the environment (described in Table 6.2)—and total acidity (Elias, 1990). Sagues et al. (2009) expanded this list to include temperature, oxygen concentration, scaling tendency (i.e., the deposition of a tightly adhered protective film of mineral solids), soil particle size distribution, porosity, and microbial activity. Although complete subsurface site characterization may not be possible for large infrastructure projects (see Chapter 5 for description of general methods of characterization), understanding subsurface variability has direct bearing on the ability to accurately model corrosion of buried steel. This chapter introduces both engineered and natural subsurface environments and describes how those environments may affect corrosion of buried steel.

SOIL

Soils, which are known among geoscientists as unconsolidated sediments, are surface and subsurface materials composed of three phases: inorganic mineral grains, void space occupied by gas (most commonly air), and void space occupied by liquids (most commonly water).[1] Box 3.1 outlines the processes by which the inorganic constituents of soil are formed. In addition to inorganic constituents, many soils contain some organic matter from decayed vegetation and active microbial communities. While some soils may be native, undisturbed in situ material, some infrastructures or environments will require excavated material used as fill (see Box 3.2 describing different types of soil). The processes of excavating, stockpiling, and filling expose soils to oxygen and can change the water content by wetting or drying. Microbial communities within the soil can be impacted by these changes,

---

[1] The definition of "soil" differs among different technical communities. Geotechnical engineers, for example, might think of soils as consolidated or unconsolidated based on their loading histories. Soil scientists might define soils as the near-surface mineral or organic layer that has undergone some type of weathering (e.g., chemical, biological, or physical). Other experts might consider soil to be that unconsolidated material on the surface that serves as a growth medium for plants. In this report, soils are unconsolidated materials at any depth.

BOX 3.1
How Is Soil Formed?

Soils typically are formed by the weathering of a parent material (e.g., rock, mineral grains) through physical, chemical, or biological processes. Physical weathering, such as freeze-thaw contraction and expansion, causes the parent material to fracture, producing smaller particles with the same mineralogy as the parent material and increased surface area. Chemical weathering occurs in the presence of water and typically produces soils with a different composition from the parent material. Chemical weathering begins with hydration and can involve some combination of hydrolysis, complexation or chelation of metals, cation exchange, oxidation, or carbonation (Mitchell and Soga, 2005). Biological weathering is the weakening and subsequent disintegration of rock by plants, animals, and microorganisms by both physical and chemical processes. Plant roots can exert physical stress on rock. Root exudates and microorganisms release chemicals that degrade rock minerals and support algae. As biodegradation processes continue, holes and gaps develop, exposing rock to additional abiotic physical and chemical weathering.

Soils are composed of a wide variety of minerals, typically ranging from quartz, feldspars, and micas (commonly encountered as sand or gravel-sized particles); to kaolinite, montmorillonite, and chlorite (which are clay minerals and exist as smaller particles); and to iron oxides and hydroxides, carbonates, and sulfates.

BOX 3.2
Types of Soils for Engineering Uses

Fill soils: excavated soils used in construction to fill depressions, alter surface topography, or engineer the ground using material types and placement procedures to result in specific engineering behaviors.

Imported soils: soil excavated for use as fill and placed in a location different from its native or original in situ location. For buried-steel applications, imported soils are normally chosen for low geochemical reactivity and free-draining properties.

Native soils: soils located in situ at a site.

Disturbed soils: soil excavated to accommodate construction and then used as backfill on-site.

Undisturbed soils: soils in their natural states, not excavated or transported.

---

as can the soil geochemistry. Compaction needs to be carefully controlled to ensure uniform density of the fill as it is being placed; intimate contact at the soil–steel interface is important because pore space at the interface can increase corrosion of the steel when those pores are subsequently filled with water (Melchers and Petersen, 2018). In buried-steel applications, imported fills are chosen primarily for their engineering properties including strength and stiffness, as well as for their corrosion-resistance properties, including low electrical activity, rapid drainage, and low geochemical reactivity. While use of fine-grained soils as fill may be unavoidable in some cases, most coarse-grained soils, including gravels and sands, provide better corrosion resistance than fine-grained soils.

The abundance of groundwater is an important defining characteristic of a soil, whether an undisturbed soil or an engineered fill. A soil is considered saturated in groundwater when there is negligible air in the void space. Below the water table, all soil voids are filled with water, and hydrostatic pressure is positive. However, full or partial saturation can still occur above the water table due to capillary rise in fine-grained soils. This rise can be on the order of meters in a clay soil. Saturation contributes to two opposing factors that affect corrosion differently. As the degree of saturation increases, the electrical conductivity increases, which tends to promote corrosion; but at the same time, lower dissolved oxygen concentrations associated with saturated conditions can inhibit corrosion by lowering the rate of the oxygen reduction (cathodic) half-cell reaction, which was introduced in Chapter 2 (see Equation 2.2). Generally, the maximum corrosion rates occur at 60–85 percent saturation (Elias, 1990). In addition to the degree of saturation, groundwater can also affect the chemistry of the soil. Dissolved constituents from the dissolution of inorganic salts, weathering of rock or soil, and the intrusion of contaminants such as road deicing salts, fertilizers, or acid mine-drainage can move into the subsurface through groundwater transport.

Soil particle or grain size is another important defining characteristic of soil. This report uses the Unified Soil Classification System (ASTM D2487-17e1, 2020) (see Table 3.1), which classifies soil according to grain size, grain size distribution, and plasticity (the nonrecoverable deformation of the soil with no cracking) (ASTM D2487-17e1, 2020). Gravels and sands are considered coarse-grained soils that display engineering behaviors governed by gravitational forces. Silts and clays are considered fine-grained soils that display engineering behaviors governed by both gravitational and electrostatic forces. Note that these definitions differ from classification schemes used in geology and agriculture. Grain size has important impacts on parameters such as electrical conductivity, because the dominant soil mineralogy changes as grain sizes decrease. The electrical conductivity of coarse-grained soils (gravels and sands) is generally low, resulting in limited corrosion; however, fine-grained soils, especially high-plasticity clays (see Table 3.1) with high cation exchange capacity (CEC), have relatively low electrical resistivity and contribute to corrosion of buried steel. The CEC of a soil is a measure of its ability to exchange or sorb positively charged ions; as CEC increases, resistivity decreases. Additionally, the organic matter present in a soil has the ability to form complexes with dissolved ions, which can increase conductivity by increasing ion solubility, binding corrosive ions, or decreasing the activity of dissolved ions through hydration. The drainage ability of a soil is also dependent on grain size and contributes to the corrosion of buried steel through the retention of water at the soil–steel interface. Coarse-grained soils can be free draining, have high hydraulic conductivity (10^{-3}–10^{-2} cm/s), and may not easily retain water at the steel surface. In contrast, fine-grained soils with low hydraulic conductivity (10^{-7}–10^{-5} cm/s) retain water at the steel–soil interface. Drainage in soils with a range of grain sizes (e.g., silty sand or clayey sands) is dominated by the smaller grain fraction that fills the space between larger grains and intermediate water retention. Additionally, the smaller particles may produce small-diameter pore networks, which tend to retain moisture through capillary action.

Two additional parameters that contribute to the advancement of corrosion are the oxidative power of the environment and pH. The corrosion cell reactions introduced in Chapter 2 include at least one oxidation reaction (metal dissolution) and one reduction reaction (usually the oxygen reduction reaction or hydrogen evolution reaction) in electrolytes such as wet soil. Other reduction reactions are also possible depending on the presence and concentration of oxidizing agents in the environment, including nitrate (NO_3^-), manganate (MnO_4^{-2}), ferric (Fe^{+3}), and sulfate (SO_2^{-4}) ions, as well as dissolved carbon dioxide gas (CO_2) (Borch et al., 2010). The reduction reactions do not consume metal like the oxidation reactions do, but they are critical in determining the rate

TABLE 3.1 Unified Soil Classifications

Unified Soil Classification	Grain Size (mm)	Secondary Descriptors
Gravel size	>4.75	Well graded, poorly graded, with silt, or with clay[a]
Sand size	0.075–4.75	Well graded, poorly graded, with silt, or with clay
Silt size	0.002–0.075	High plasticity or low plasticity[b]
Clay size	<0.002	High plasticity or low plasticity

[a] A well-graded soil has a wide range of particle sizes, with no concentration of particles in a single size and no gaps in the grain size distribution. A poorly graded soil consists of particles that are predominantly one size, as defined in ASTM D422-63 (2016). Note: In the geosciences, a well-graded soil is defined as "poorly sorted" and a poorly graded soil is defined as "well sorted."

[b] Plasticity is the ability of a soil to deform without cracking, and is quantified as the range of water content between the soil's liquid limit (water content at which the soil flows like a liquid) and plastic limit (water content at which the soil exhibits cracking under applied stress) as defined in ASTM D4318-17e1 (2018)

SOURCE: Adapted from ASTM D2487-17e1 (2020).

of metal corrosion, which is often limited by the rate of transport of the cathodic reactant or electron acceptor to the steel surface. In many environments, dissolved oxygen is the energetically preferred electron acceptor. This is often the case in high hydraulic conductivity environments, where dissolved oxygen can easily be replenished via groundwater transport. In other environments (e.g., low hydraulic conductivity environments), oxygen cannot easily be replenished. Unless some of the other electron acceptors mentioned above are present, these environments will have limited oxidizing power (i.e., redox potential). The rate, or kinetics, of an electrochemical reaction often depends exponentially on the electrochemical potential, which is influenced by the oxidizing power of the environment, whereas the thermodynamic tendency for the reaction to proceed depends on both potential and concentration of the reactants and products.

Because many of the important oxidation and reduction reactions involve hydrogen (H^+) or hydroxyl (OH^-) ions, pH is important. The effects of potential and pH are captured by Pourbaix diagrams, which are plots in potential (E)-pH space of thermodynamic equilibrium conditions for reactions in a corroding metal system (Pourbaix, 1974). Figure 3.1 is a Pourbaix diagram for iron (Grundl et al., 2011). The text shows the E-pH field of predominance for each species. The lines represent the conditions for equilibrium between the two phases that the line separates. For example, the horizontal line at the bottom of the plot is for the equilibrium between solid metallic

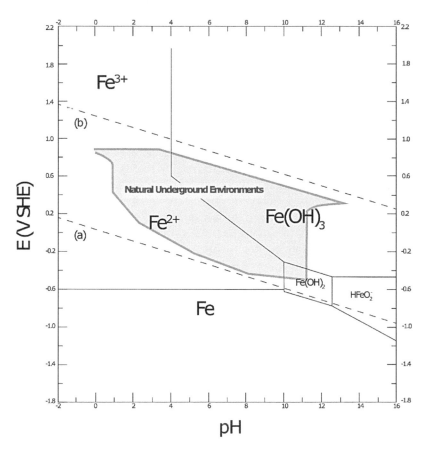

FIGURE 3.1 Pourbaix diagram for iron. The *x*-axis represents the pH of the environment. The *y*-axis represents the electrochemical potential, which is a measure of the oxidizing power of the environment. The solid lines represent equilibrium conditions for the reactions between the various species. The dotted lines are found on all Pourbaix diagrams and represent the equilibrium potentials for the hydrogen evolution (a) and oxygen reduction (b) reactions. These cathodic reactions will take place at potentials below the respective lines and drive corrosion of iron. The filled gray shape represents the region of most natural underground environments.
SOURCE: Modified from Grundl et al. (2011).

iron (Fe) and ferrous (Fe^{2+}) ions. Thermodynamic considerations indicate that metallic iron will either corrode to form Fe^{2+} or react to form an oxide at regions of potential and pH above these lines. The gray area surrounded by the bold line represents the *E*-pH range of natural underground environments. This region is located above the region where metallic iron is stable, and so iron and steel will be oxidized in these environments.

Corrosion will be slow at relatively high pH values where a solid oxide such as iron(II) hydroxide ($Fe(OH)_2$) or iron(III) hydroxide ($Fe(OH)_3$) is stable, because the spontaneous formation of the oxides can provide protection (this probability of oxide formation is called "scaling tendency"). However, steel will corrode extensively in neutral and acidic environments (e.g., pH <4.5; Roberge, 2000; Shreir et al., 1994), which is common in subsurface conditions. While the pH of water in equilibrium with the atmosphere is approximately 5.7 because of the effects of carbon dioxide in air, the pH of the subsurface is often controlled by the minerals in the surrounding soil and rock (e.g., pH of clay and iron-bearing soils tends to be less than 7; pH of carbonate-containing soils tends to be greater than 7). Additionally, respiration, especially aerobic respiration, will add carbon dioxide. Natural organic matter, which consists of highly complex chemical compounds of carbon, hydrogen, oxygen, nitrogen, phosphorus, and sulfur resulting from the degradation of organic materials (e.g., vegetation, microbes), may release H^+ ions that decrease the pH.

ROCK ENVIRONMENTS

Rocks are solid masses or cemented aggregates of minerals and glassy noncrystalline materials (e.g., volcanic glasses). They are broadly classified as igneous (forming through cooling and solidifying of magma), metamorphic (alterations of preexisting rock through changes in temperature and pressure), or sedimentary (cementation of rock and mineral fragments after deposition). Table 3.2 lists some commonly encountered rock types. Grain size in rocks refers to the size of the individual mineral grains or crystals that make up the rock mass, with sizes that vary from microscopic to centimeter or larger scale. Rocks may also include void space at as much as 40 percent of total volume, which is a function of the formation environment. Voids in rocks can be occluded and inaccessible to fluid flow, or open to flow and permeation by gases and liquids. Joints, faults, bedding planes, and other discontinuities in a rock mass typically impact the engineering behavior of the rock mass and control the transport of fluids (water, gases, and other liquids) through the rock mass, creating zones of differential moisture and chemical and oxygen contents and affecting corrosivity within those zones. Rock and steel are seldom in direct permanent contact, but the tip of a steel foundation pile, for example, may be driven through soil to be end-bearing on rock. In all other instances, except some temporary supports, steel buried within a rock mass is installed in an oversized hole that is then backfilled with cementitious grout or concrete, a chemical resin, or soil.

ENGINEERED FILLS

It is often the case that native soils or rock are unable to provide the conditions necessary to support the performance of infrastructure. In such cases, engineered fills may be used. Engineered fills are soils or granular materials that are treated or created with specified particle size, chemistry, moisture content, or plasticity. They are often mechanically placed in layers and compacted to specified density to achieve performance goals specified by an engineer. They can be designed or selected to be less corrosive than the native subsurface. However, corrosion can still occur in engineered fill due to its inherent properties or a change in environment.

Engineered fills may be in contact with steel as backfill (engineered fill placed in an excavation) around pipes and storage tanks, or as part of the material penetrated by a steel pile for a foundation or to retain an excavation. A common occurrence of steel buried in engineered fill is in mechanically stabilized earth construction.

TABLE 3.2 Commonly Encountered Rock Types

Igneous	Metamorphic	Sedimentary
Diabase, diorite, granite, basalt, obsidian, tuff	Gneiss, schist, slate, marble	Conglomerates, sandstones, shales, limestone

GROUT, CONCRETE, AND FLOWABLE FILL

The subsurface environment around buried steel may include grout, concrete, or flowable fill. Grouts solidify after application but upon application can be categorized as suspensions, solid particles suspended in a liquid phase (e.g., cementitious grouts); emulsions, liquid phase suspended in a liquid phase (e.g., bitumen in water); foams (e.g., gas bubbles in a liquid); and solutions, molecular mixtures of two substances (e.g., chemical grouts). These are discussed the next section. The ability of grout to flow is determined by its viscosity, which is controlled by variables such as ratio of water to cement, the addition of superplasticizers, and additives that affect viscosity and slow the cure time.

Cementitious grout is a grout with various proportions of cement and water. It may include other constituents, such as clay (e.g., montmorillonite), to achieve desired properties. Cementitious grout is used to permeate soils and to fill rock fractures, and its most common use with respect to buried steel is to surround steel inserted in a drill hole (e.g., ground anchors). The grout suspension is pumped into the annular space of a drill hole, between the steel and soil or rock, where it cures to provide a bond between the steel and the earth material. Concrete is similar to cementitious grout but with sand and gravel additives of various sizes, called aggregates. Corrosion of steel bar in reinforced concrete is an important process that has received considerable study (e.g., Broomfield, 2003), but it is not considered in this report. Similar to concrete, flowable fills also use cement and aggregates but in different proportions such that they are less viscous and often more permeable. The cured strength is typically closer to that of soil than concrete.

During hardening of cement mixtures, water is generally lost to the environment, leaving voids that can transmit fluids. The permeability created by these newly formed voids provides an environment in which corrosion mechanisms might be initiated. Subsequent applied loads, thermal cycles, and wetting and drying cycles will further increase permeability via microcracking. Mehta and Monteiro (2014) outline several subsurface processes and reactions that are particularly damaging to cement mixtures and thus affect permeability. These are often the result of or triggered by the presence of water. For example, gels form when acidic siliceous soils react with alkaline concrete pore fluids. The gels expand in the presence of water to cause cracking (alkali-silica reactivity). Water in pore spaces may simply expand when frozen and contract when thawed (freeze-thaw cycles), causing cracking. Cracking in cements may also be caused by a series of chemical reactions that occur between cement paste and sulfate ions (sulfate attack).

Noncementitious grouts may be made of silicates, acrylics, and polyurethanes. Epoxy resin grouts are often used to grout steel into drill holes. The epoxy resin is similar to that used explicitly for corrosion protection on steel (see Chapter 5), so the steel is well protected where coverage is complete. However, since mixing of the resin and its distribution throughout a drill hole occurs in the field, the coverage is often imperfect and corrosion can occur at gaps in coverage.

4

Corrosion of Buried Steel

As described in Chapter 2, metallic corrosion involves oxidation (anodic) and reduction (cathodic) half-cell reactions, where the metal oxidation reaction is the anodic reaction, and the reduction of a species in the subsurface (e.g., oxygen, water, or H^+ ions) is the cathodic reaction. Although the electrochemical fundamentals of metallic corrosion in the subsurface are always the same, the corroded component might look different under different conditions because there are several forms of corrosion. Corroded steel might exhibit uniform attack over its entire surface, or the corrosion might be localized at a few spots on the steel. Alternatively, the steel might exhibit little change on the surface, but the absorption of hydrogen might make the steel brittle. This chapter provides details regarding the fundamentals of buried steel corrosion, and different forms of corrosion are addressed.

GENERAL CORROSION

Figure 4.1 illustrates the reactions occurring at anodic and cathodic sites on a buried steel component. Iron dissolution occurs at anodic sites, and reduction reactions occur at closely located cathodic sites. Electrons flow between them. The anodic and cathodic sites on the steel surface often will move continually across the steel surface such that all parts of the surface eventually experience the anodic dissolution reaction and dissolve (i.e., corrode). The resulting uniform wastage of material across the surface is known as general or uniform corrosion. The corrosion rates of general corrosion can be estimated by the rate of loss of steel thickness or by the rate of weight loss normalized by the exposed area. Buried steel commonly corrodes by a general corrosion mechanism, although the attack might be localized in certain spots known as pits (Romanoff, 1964). General corrosion will often be present along the surface of buried steel components unless the surface is coated with a dielectric, or an electrically insulating, material. The severity of general corrosion depends on the soil conditions (e.g., moisture, degree of aeration, pH). The corrosion can be so minimal as to be insignificant, or it can be severe with a significant impact on the durability and service life of the steel. Typical general corrosion rates in soil result in a loss of about 0.045 millimeters (mm), with the deepest pits of about 0.23 mm over a period of 9 years (Shreir et al., 1994).

As described in Chapter 3, the oxidizing power of the underground environment, which is controlled by the degree of aeration and the presence of other oxidizing species, will impact the corrosion potential. The oxidizing power can have a large effect on the rate of general corrosion because the corrosion rate increases exponentially with an increase in electrochemical potential. The equation describing this relationship is called the Tafel equation. The Tafel slope is the slope of the semilogarithmic plot of potential versus log of the current density (a measure of

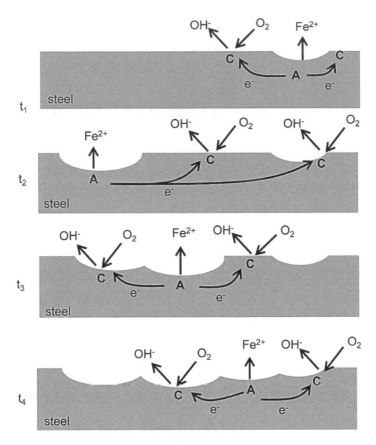

FIGURE 4.1 Schematic representation of general corrosion showing different spatial relationships between anodic (represented as "A" in the figure) and cathodic (represented as "C" in the figure) sites. Anodic sites are where iron ions (Fe^{2+}) are released (i.e., corrosion). Progression of corrosion over times t_1 through t_4, where the anodic and cathodic sites move across the surface with time, resulting in nominally uniform loss of the steel.

rate) (see Figure 4.2). The fundamental understanding of corrosion kinetics as expressed by the Tafel equation can be used in an accurate determination of corrosion rate (corrosion current density) using different electrochemical methods. The kinetics of the cathodic reaction often also exhibit exponential dependence of current density on potential (i.e., the Tafel equation but with a different Tafel slope). The curve representing the cathodic oxygen reduction reaction (see Equation 2.2) is often a vertical line, which indicates that the rate of the reaction is limited by the transport of oxygen to the steel surface by diffusion. In this case, the rate of steel corrosion can be controlled totally by the rate of transport of the cathodic reactant (i.e., dissolved oxygen) to the metal surface. In liquids, transport is accelerated by convection, even natural convection that occurs in the absence of any forced flow; however, convection is limited in soils with sand-size and smaller particles due to reduced flow rates as water flows through the small tortuous paths in pore spaces between. As a result, transport of oxygen to the surface of buried steel might be diffusion limited and occur at a low rate.

The temperature of the environment can affect the corrosion rate in multiple ways. The electrochemical reactions represented by Equations 2.1–2.5 are thermally activated, meaning that their rate will increase, often exponentially, with increasing temperature. Therefore, corrosion rate tends to increase with increasing temperature. Furthermore, diffusion is also a thermally activated process, and so for the case of corrosion limited by oxygen diffusion, the corrosion rate will also increase with increasing temperature. However, the solubility of oxygen in water decreases with increasing temperature, and this will tend to decrease the rate of corrosion. For steel corroding in liquids, these factors balance to result in a maximum corrosion rate at 70°C.

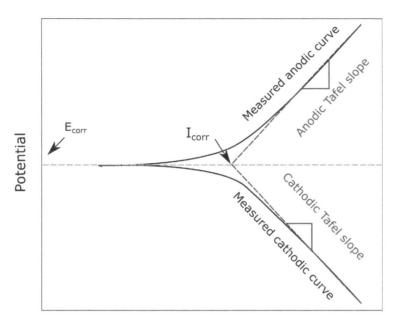

Logarithmic current density

FIGURE 4.2 Idealized potential–current density plot. Corrosion potential E_{corr} is where the curve points to the left, the anodic and cathodic Tafel slopes are the linear parts of the curve far from E_{corr}, and the corrosion rate measured by the corrosion current density I_{corr} is where the extrapolated Tafel lines intersect E_{corr}. A typical corrosion potential of steel in soil is –500 mV to –750 mV versus a copper–copper sulfate reference electrode (with no cathodic protection), and a typical corrosion current density is on the order of $2E\text{-}6$ A/cm^2.

LOCALIZED CORROSION MECHANISMS FOR BURIED STEEL

As shown in Figure 4.1, anodic and cathodic sites can continually move across the surface with time, causing general corrosion or uniform steel loss. However, when the anodic and cathodic sites are spatially fixed (see Figures 4.3–4.7), the corrosion will be localized at the fixed anodic sites. Such localized forms of corrosion are described in the next sections, and it is important to distinguish between general and localized corrosion because of the structural implications related to targeted attack. The same fundamental electrochemical mechanisms control the corrosion, but each form of corrosion has a different appearance. One distinguishing difference is the spatial separation of the anodic and cathodic sites. For general corrosion, in which the anodic and cathodic sites continuously move, their separation distance might be on the order of nanometers. The separation for localized corrosion can be micrometers (pitting corrosion), millimeters (crevice corrosion), or centimeters to meters (macrocell corrosion).[1] In buried-steel applications, general corrosion, pitting, and macrocell corrosion are generally the corrosion mechanisms that need to be mitigated or accounted for during design or analysis of performance. Other forms of corrosion including galvanic corrosion and crevice corrosion can be avoided with proper design. Mechanisms for localized corrosion methods are described in the next sections.

---

[1] Terminology and scales associated with localized corrosion had to be defined by the committee through research and deliberation. For example, it took some time for committee members to realize that the corrosion engineers and metallurgists on the committee defined the pitting corrosion and its scales differently than the geotechnical and civil engineers on the committee.

Pitting Corrosion

Pitting is localized corrosion that results when the anodic site is spatially fixed on a boldly exposed surface (see Figure 4.3) as opposed to within an occluded crevice. The separation distance between anodic and cathodic sites is on the order of micrometers to millimeters. The anodic reaction inside the pits is supported by cathodic reaction on the nearby surface. Pits usually initiate at susceptible sites on the surface associated with metallurgical features. Once a pit initiates, it tends to propagate because oxygen is depleted inside the pit while being readily available on the outer surface. This allows continued cathodic oxygen reduction reactions only on the exposed surface. A single deep pit in pipelines can initiate a leak that is large enough to require immediate repair and response. On the other hand, for structural systems where containment is not one of the functions, a deep pit may not affect performance and service life.

Aboveground pitting is commonly associated with a surface that is protected by a thin oxide film called a passive film. Plain carbon steels commonly used in underground environments may form a protective iron oxide layer if the local pH is relatively high. As the oxide layer begins to break down, for example, through reactions with dissolved ions such as chlorides, the steel becomes susceptible to pitting corrosion. Cations generated by the pitting process react with water causing strong acidification within a pit, and aggressive anions such as chloride are enriched in a pit by a process called migration. As a result, the environment in a pit often resembles hydrochloric acid, creating an aggressive environment that promotes sustained dissolution in the pit. However, most underground environments have pHs in which these passive films do not generally form on plain carbon. The pits commonly observed in buried steel can be better classified as a form of microcell corrosion as described below. Pits in buried steel can also result from microbially influenced corrosion (MIC), as also described below.

Multiple converging pits, hemispherical pits with interior pitting, multiple tiers within pits, striations within pits, and tunneling morphologies have all been identified in case studies of MIC involving carbon steel (Chen et al., 2021; Islam et al., 2016; Pope, 1990). As mentioned above, local pit environments are often acidified by reactions of dissolved cations with water. Pits formed by MIC are similarly acidified but are also influenced by microbial metabolic processes. In their review of sulfate-reducing bacteria (SRB) corrosion, Enning and Garrelfs (2014) included images of localized corrosion on the exterior of a buried carbon steel pipe exhumed from a water-logged, anoxic, sulfate-rich (1.4 mm) soil (see Figure 4.4). The images are consistent with previous descriptions of MIC of carbon steel.

Crevice Corrosion

Crevice corrosion (see Figure 4.5) occurs in occluded regions formed by the close contact of a metal and a nonmetal (polymer or ceramic) or another piece of the same metal with cathodic-anodic spatial separation on

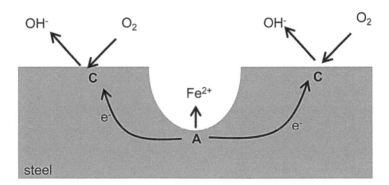

FIGURE 4.3 Schematic representation of pitting corrosion showing different spatial relationships between anodic ("A" in the figure) and cathodic ("C" in the figure) sites. In this case, the anodic site does not move, and the dissolution remains focused at a location that deepens with time. The outer surface is usually protected by a thin oxide passive film, and it corrodes slowly in comparison to the pit.

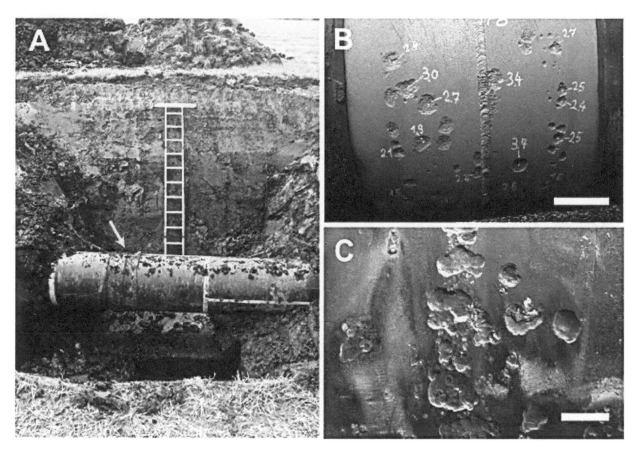

FIGURE 4.4 (**A**) External corrosion on buried gas transmission pipeline excavated from a water-logged, anoxic, sulfate-rich soil. (**B and C**) Details of corrosion under a disbonded asphalt coating, illustrating clusters of pits and pits within pits. Numbers in (**B**) denote pit depths in mm and the bar is 20 cm. The bar in (**C**) is 2 cm.
SOURCE: Enning and Garrelfs (2014).

the order of millimeters. One example of this is the connection of a structural element (e.g., nuts, bolts, washers, bolt holes). Local contact with rock might also be the site of crevice corrosion. The occlusion causes a physical separation of anodic and cathodic sites because of the limited supply of oxygen (the common cathodic reactant) in the crevice. This spatial separation leads to the same development of a corrosive local environment as occurs in a pit. In fact, a deep pit is a crevice. The difference between pitting and crevice corrosion is that crevice corrosion will initiate more easily than pits. If the entire surface of a buried steel structure is not completely covered with a protective coating, crevice corrosion can initiate where there is a small gap between the coating and the metal. Crevice corrosion can be avoided by sealing crevices, applying cathodic protection (CP), or using more corrosion-resistant materials.

Galvanic Corrosion

Galvanic corrosion (see Figure 4.6) occurs when dissimilar metals are in electrical contact in the same electrochemical environment. The metal that has the more negative electrode corrosion potential (i.e., it is less noble) becomes the dominant anodic site and corrodes more quickly than when not electrically connected. The metal with less negative corrosion potential (i.e., more noble) becomes the dominant cathodic site and corrodes more slowly than when not electrically connected. The relative nobility of metals is indicated by the galvanic series, which lists the corrosion potential of metals from relatively active metals such as zinc to relatively noble metals

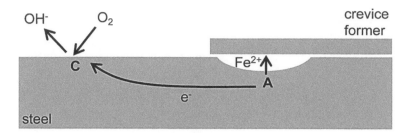

FIGURE 4.5 Schematic representation of crevice corrosion showing different spatial relationships between anodic ("A" in the figure) and cathodic ("C" in the figure) sites. In this case, crevice corrosion is occurring under a crevice former.

such as copper. Figure 4.6 shows an example of the galvanic coupling of steel with brass, which would result in accelerated corrosion of the steel near the contact point with the brass.

Galvanic corrosion may occur when bolt materials differ from the material used in the infrastructure component (e.g., plain steel bolts on sections of stainless steel or weld materials that differ in composition from the base steel). Galvanic corrosion can also occur when galvanized (i.e., coated with zinc) and plain steel are in contact (e.g., when a nongalvanized temporary structure comes in contact with permanent galvanized reinforcements or when a nongalvanized reinforcing steel and galvanized tie strips come in contact in a concrete facing wall). Galvanic corrosion is not usually considered a localized form of corrosion but is included in this section because the lateral extent of the galvanic interaction is often limited to the region close to the dissimilar metals' connection, as is shown in Figure 4.6.

There are examples wherein galvanic corrosion is used to the engineer's advantage and contributes to corrosion management practices. For example, coating steel with zinc creates a galvanic couple between the zinc and steel surface. The zinc will corrode preferentially to the steel and the underlying steel will not be consumed until the zinc coating is exhausted, thus affording corrosion protection to the steel.

Macrocell Corrosion

In a variety of steel structures including piles, soil nails, or pipelines (see Table 2.1 for descriptions), localized regions of the steel may be exposed to greater levels of oxygen than other regions. Those aerobic (oxygen-rich) regions would be the preferential cathodic sites, and the oxygen-poor regions would be preferential anodic sites. In some cases, such as retaining-wall systems that use ground anchors to support a steel facing, current may travel through millimeters to meters of steel from the anodic site (e.g., the bonded zone of the anchor) to the cathodic sites (e.g., the steel face). An area of steel serving as a net anode in this reaction will corrode at a higher rate than an area serving as a net cathode. Figure 4.7 represents a pile that is driven through different soil types, specifically a loose sand layer over a dense clay. The less-dense soil (i.e., the loose sand) allows greater access to oxygen and is the site of the cathodic reaction, with corrosion focused in the region with less oxygen.

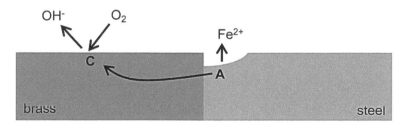

FIGURE 4.6 Schematic representation of galvanic corrosion showing different spatial relationships between anodic ("A" in the figure) and cathodic ("C" in the figure) sites. In this case, galvanic corrosion is occurring between steel and brass.

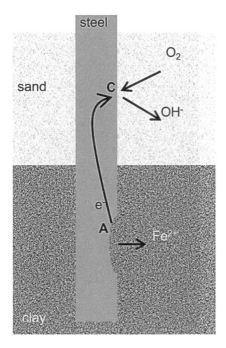

FIGURE 4.7 Schematic representation of macrocell corrosion showing different spatial relationships between anodic ("A" in the figure) and cathodic ("C" in the figure) sites. In this case, a macrocell was formed by a pile driven through a heterogeneous soil consisting of a loose sand layer over a dense clay layer. Corrosion is focused in the region with less access to oxygen.

Macrocell corrosion driven by differences in oxygen concentration can also be referred to as "concentration cell corrosion" or "differential aeration corrosion." This type of corrosion might result in localized attack of the steel to form a region often referred to as a pit. However, this pitting of buried steel is different than the pits formed on passive steel that might occur in high-pH soils containing chloride ions, as described above. The latter situation, which is rare, would exhibit small spots of rapid attack on a surface that is otherwise relatively unattacked. The more common pitting of buried steel, resulting from microcell corrosion, takes the form of heavy corrosion across the component, with some sites exhibiting more rapid thinning. The rate of this type of pitting is often greater in poorly aerated soils (Romanoff, 1957).

Macrocell corrosion can occur, for example, in drilled shafts where concrete covers the reinforcing steel (see Table 2.1) or when the cover is breached from voids or a soil inclusion at the edges of the concrete. A macrocell is created between the portion of the bar that is embedded within concrete and the portion exposed with the void space. Galvanic currents driven by dissimilar media like this can cause the corrosion rate of exposed steel to increase by 3.3 to 5.6 times (Sarhan et al., 2002).

Stray-Current Corrosion

Stray-current corrosion occurs when buried steel inadvertently interacts with nearby sources of alternating current or direct current (DC) voltage. One common producer of stray currents is DC-powered transit or other rail systems (Sankey and Hutchinson, 2011). Stray electrical currents may also exist around electrical transmission systems, waterfront structures in saltwater, CP systems, or welding shops. They often encounter buried metallic structures including buried utility pipes and cables, underground storage vessels, and reinforced concrete structures. This type of corrosion is most commonly observed on structures that have large dimensions in one direction such as pipelines, and, in fact, much of the experience with stray-current corrosion is from the observed performance of pipelines. Sheet piling and other piling that are electrically continuous also can experience stray-current corrosion (Beavers and Durr, 1998). In particular, buried metallic structures in high-density urban areas are at risk.

CP is a means of protecting buried steel by making it the cathode of an electrochemical cell and is described in more detail in Chapter 5. Figure 4.8 shows a schematic for stray-current corrosion associated with CP, where a DC power supply passes cathodic current to a steel tank from a remote anode to reduce the corrosion rate of the tank. In this diagram, a pipeline is unknowingly situated near the tank. Electrical current will always take the path of least resistance, which is often a straight line. However, because the electrical resistivity of the steel pipe is much less than the soil, the lowest resistance follows a path through the pipe. The current enters the pipe at a location near the CP anode, resulting in a localized cathode where oxygen is reduced (cathodic reaction), passes through the pipe as electrical current, and then exits the pipe at a location near the tank, resulting in a localized anode where metallic iron of the steel pipe is oxidized (anodic reaction).

ENVIRONMENTALLY INDUCED CRACKING

Environmentally induced cracking (EIC) involves the synergistic interactions of stress and a corrosive environment (Fletcher, 2005). There are generally three types of EIC: hydrogen embrittlement, stress corrosion cracking (SCC), and corrosion fatigue. Note that the terminology can be confusing because some experts consider hydrogen embrittlement to be a mechanism of SCC.

Hydrogen can be generated through the cathodic hydrogen evolution reaction that accompanies corrosion in most anoxic (i.e., without oxygen) environments (Equations 2.4 and 2.5). Hydrogen atoms are very small and may be absorbed into the corroding steel in relatively high concentrations, and given the volume of material involved, can cause hydrogen blistering (see Figure 4.9). Additionally, high-strength steel can become brittle due to absorbed hydrogen, leading to possible catastrophic failure under load. This phenomenon is referred to as hydrogen embrittlement.

SCC involves the synergistic interactions of stress and a corrosive environment. In fact, three factors must be present for SCC: tensile stress, a susceptible material, and a particular environment that promotes SCC of the material. SCC is prevented by the elimination of one of those three factors. SCC of pipelines can occur from the inside because of corrosive products being transported, but this is not within the scope of this report. Exterior SCC of pipelines has been observed under disbonded coatings. For example, coal tar coatings are often susceptible when the local pH and potential have been altered by CP (coatings and CP are described in Chapters 4 and 5) (Beavers, 2014). Higher temperature increases susceptibility, and a sufficiently high pH or low potential can prevent SCC. Stresses in pipelines can arise from internal pressurization, residual stresses at welds, or bending from movement

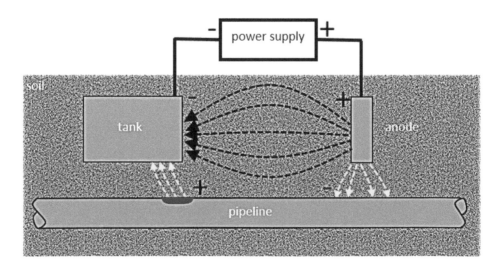

FIGURE 4.8 Schematic of stray-current corrosion. In this case, some of the current flowing in a cathodic protection system flows through a nearby pipeline. The dashed lines represent ionic current in the soil, and the yellow dashed lines represent the stray current. Enhanced corrosion occurs where the current exits the pipeline.

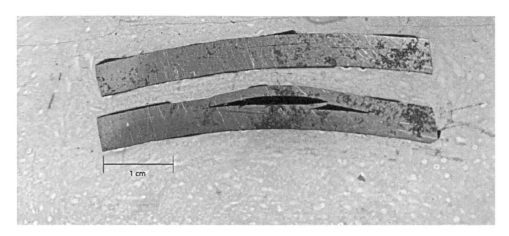

FIGURE 4.9 Example of hydrogen blister in small-diameter pipe wall.
SOURCE: Photo by Elizabeth Rutherford, committee member.

of the supporting soil. A different type of SCC in pipelines has been observed under coatings that are imperme-able to CP (e.g., polyethylene tape) if water from the environment can reach the pipe surface. The nature of the cracking in this case is different in that it is transgranular instead of intergranular.

Steel components in civil structures such as rock bolts are not susceptible to the same forms of cracking as pipelines because CP is not used. However, cracking from hydrogen embrittlement is possible for high-strength steel structures (Grade 150 or higher) when the element is subject to tensile stress (e.g., prestress) exceeding 50 percent of its ultimate tensile strength and the element is in direct contact with the subsurface environment (i.e., subject to chlorides or free hydrogen) (Fédération Internationale de la Précontrainte, 1986; Witham et al., 2002). SCC is mitigated by incorporating corrosion protection in the design of those metal tensioned elements such that the steel tendons are isolated from the underground environment. For example, isolation may be achieved with a system that includes coating the steel with grease, and covering it with a plastic sheath that is then surrounded by grout (Post-Tensioning Institute, 2014). In this case, service life is dependent on the workmanship of the installa-tion and the durability of the materials used in the corrosion protection system.

Time-varying stresses can cause cracking of metals by a phenomenon called metal fatigue even when the stresses are relatively low. The cracking is faster in a corrosive environment, which is a form of corrosion called corrosion fatigue. The loads on buried structures are usually invariant, although varying internal pipe pressure can cause a variable hoop stress that can accelerate the crack growth rate (Beavers, 2014).

MICROBIALLY INFLUENCED CORROSION

MIC of buried steel is the result of microbial activities, typically within biofilms that are bound to the sur-face of buried steel (see Chapter 3 for a description of microorganisms in the subsurface). The term MIC does not denote a specific mechanism for corrosion but rather refers to the microbial activities that can create a more corrosive environment. Microbial activities that can influence the corrosion of steel include (1) conversion of a protective iron oxide to a less protective sulfide by sulfide-producing procaryotes (SPPs), (2) direct removal of a protective oxide layer by iron-reducing bacteria (IRB), (3) direct removal of a protective oxide layer by acid-producing bacteria (APB), and (4) direct uptake of electrons from iron. Specific consequences are determined by reactions within biofilms.

Microorganisms are commonly classified as aerobes and anaerobes, depending on whether they respire oxygen. Aerobic microorganisms respire by using oxygen as the electron acceptor. Anaerobic microbes use an electron acceptor other than oxygen (e.g., nitrate, sulfate, ferric, or manganese ions) in anaerobic respiration. A facultative anaerobe is an organism that uses aerobic respiration if oxygen is available but can switch to other electron acceptors if oxygen is absent.

Temperature can affect many aspects of MIC. Temperature determines the growth rate and distribution of specific microorganisms, as well as the reaction rates of any microbial metabolites (e.g., acids and sulfides) with carbon steel. Microbial species have been identified at temperatures ranging from −10°C to more than 100°C. However, each microbial species has an optimum temperature range for growth and a maximum temperature for survivability. Both are determined by cell metabolism. In general, within the optimum temperature range for growth, increases in temperature increase metabolism and growth rate. The impact of temperature on dissolved oxygen means that decreases in dissolved oxygen will influence the distribution and extent of aerobic respiration.

Microorganisms typically associated with MIC are listed below based on specific metabolisms and corrosion-causing mechanisms (e.g., sulfide production, iron reduction, acid production, and methanogenisis). Vigneron et al. (2018) concluded that such classifications of MIC-related microorganisms by a single metabolic function can be misleading. Microorganisms associated with MIC are metabolically versatile and capable of expressing multiple pathways other than those commonly attributed to them. Corrosive biofilms contain numerous microhabitats with different redox potentials and chemical gradients, allowing the establishment of microorganisms with parallel, complementary, and antagonistic physiologies (Vigneron et al., 2018) that are rarely acknowledged. In the following sections, known complementary syntrophic interactions within biofilms, where one microbial species lives off the products of another species, have been included.

Although MIC is not considered a specific mechanism of corrosion, it is important for designers, metallurgists, and users of buried steel structures to understand the basic mechanisms and the conditions in which this type of corrosion can occur. These are described in the following sections. MIC is generally not taught in university curricula covering corrosion of steel. Because of the growing importance and recognition of MIC (see Box 4.1), metallurgical and materials curricula would be well served to include the study of MIC as a critical subject of corrosion.

Sulfide-Producing Procaryotes

Under anoxic conditions, steel is highly susceptible to MIC. SRB are the group of anaerobic bacteria most closely identified with sulfide production because of sulfate reduction. SRB activity reportedly causes the average corrosion rate of steel exposed to soil in the absence of oxygen to be >20 times higher than that of abiotic controls (Li et al., 2001). Most published reports of MIC on cathodically polarized steel are attributed to SRB, sometimes in association with APB, under disbonded coatings (Abedi et al., 2007; Li et al., 2000). SRB oxidize specific electron donors (e.g., molecular hydrogen, methanol, ethanol, acetate, lactate, propionate, butyrate) by reduction of inorganic sulfate (electron acceptor) to sulfide. Iron oxide, often found on the surface of buried steel, readily reacts with the sulfide. Non-SRB can produce sulfide by reducing other oxidized forms of sulfur (e.g., thiosulfate, sulfite, or green rust class 2 [$GR2(SO_4^{2-})$]). Many archaea can also produce sulfides. The inclusive term for sulfide-producing anaerobes is SPPs. As more sulfides are produced by SPP, the sulfide-deficient iron corrosion product (mackinawite, FeS_{1-x}) is converted to a sulfide-rich mineral (e.g., greigite, Fe_3S_4). Accumulation of microbiologically produced iron sulfides on iron surfaces stimulates the cathodic reaction. Once electrical contact is established, a galvanic couple develops with the steel surface as an anode and electron transfer through the cathodic iron sulfide. Introduction of oxygen causes conversion of the sulfide back to an oxide and an immediate increase in the corrosion rate (Blackwood, 2020).

Some SRB have been identified as electrogenic microorganisms, capable of transferring electrons to or from solids. Venzlaff et al. (2013) concluded that some "specially adapted, highly corrosive SRB" derived energy directly from elemental iron (Fe^0). Their observations led to the conclusion that cells starved of organic carbon were more aggressive to steel. However, it is unclear how many SRB can use iron as an electron donor. Enning and Garrelfs (2014), for example, concluded that only a few SRB are capable of electrogenic reactions, whereas, Y. Li et al. (2018) suggested that Fe^0 can more generally be an electron donor for SRB when there is a lack of carbon sources.

In laboratory experiments, Chen et al. (2021) suggested that multiple tiers within pits were indicative of SRB-influenced MIC of API X80 steel pipe. Specimens were examined after 14-day exposures to sterile and inoculated (*Desulfovibrio desulfuricans*) soil suspensions. Localized corrosion was observed in both abiotic and biotic exposures (see Figure 4.10). The authors concluded that the amounts, sizes, and depths of pits significantly

BOX 4.1
Case Study: Microbially Influenced Corrosion of the
Leo Frigo Memorial Bridge, Wisconsin

A severe steel pile corrosion event occurred at the Leo Frigo Memorial Bridge near Green Bay, Wisconsin, on September 25, 2013. Between 3:00 a.m. and 3:45 a.m., Pier 22 moved vertically down 2 feet due to corrosion of the steel piles. This created an obvious "dip" in the roadway (see Figure 4.1.1 a), which was reported during the middle of the night by a truck driver on the bridge. This settlement of Pier 22 could have led to catastrophic failure of the main span, but fortunately the steel girder framing was robust enough to allow for redistribution of load to other neighboring piers. The bridge was closed to traffic, creating a major disruption to vehicular movement east to west in Green Bay. An investigation ensued and repair strategies for remediation were developed. The initial investigation relied on structural and geotechnical engineers to assess the possible cause of the settlement. Structural engineers made sure that the bridge was safe enough to implement repairs, and geotechnical engineers conducted subsurface investigation to find the source of the settlement.

Once the source of the failure was discovered to be some form of corrosion within the uppermost level of the fill material, and because of the proximity of the Fox River to Pier 22, it was concluded that the fill material was not native and that there were obvious signs of water-level fluctuations in the river. A hydrogeologist was engaged to assess possible historical fluctuations of water near this pier to gain an understanding of possible corrosion mechanisms including anaerobic and aerobic forms. Additionally, because there was a cathodically protected gas pipeline in the area, a cathodic-protection specialist was brought in to assess and measure whether stray currents could have contributed to the observed corrosion. Measurements indicated that no stray currents were in the vicinity of the bridge piers.

Aggressive cleaning of the failed steel pile surface showed corrosion pitting morphology that is consistent with microbially influenced corrosion (MIC). Samples of the fill material and underlying native soils were tested in the laboratory to type present microbes and conduct chemical analysis of the fill materials and native soil. Testing confirmed the presence of significant levels of sulfate-reducing bacteria and acid-producing bacteria in the fill. MIC-induced pitting was likely caused by the observed aerobic and anaerobic bacteria that were growing on shredded wood. The investigation also identified moist, porous, fly ash fill over oxygen-poor, low-permeability clay, which together formed a macrocell (WisDOT, 2015).

Other piers near Pier 22 were assessed, and four other piers had signs of MIC and required remediation (see Figures 4.1.1 b and c). The remainder of the 51 piers had only nominal signs of uniform corrosion that would be expected for steel piles in service for about 40 years. Both structural and geotechnical engineers, with the assistance of a corrosion engineer, developed the remediation of the five piers, which included specific corrosion protection measures that would not require the removal of the suspect fill material. The investigation and remediation of the Leo Frigo Memorial Bridge required the specialty knowledge of structural engineers, geotechnical engineers, a hydrogeologist, a cathodic-protection specialist, and corrosion engineers, and specialty testing for the presence, amount, and kind of microbes and chemical analysis of fill and soil materials. The bridge was reopened in January 2014, 4 months after discovery of "the dip."

FIGURE 4.1.1 (a) Leo Frigo Memorial Bridge and **(b)** an example of a pristine pile at Leo Frigo Memorial Bridge. **(c)** Deflection of the bridge pier was caused by pile corrosion from microbially influenced corrosion, as well as macrocell formation within a layer of fly ash fill material just below the concrete pile cap.
SOURCE: © Jim Matthews—USA TODAY NETWORK.

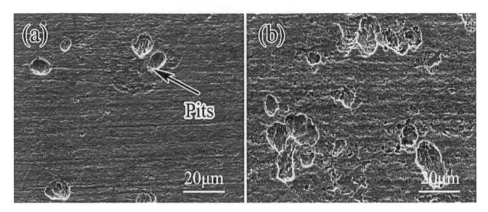

FIGURE 4.10 Scanning electron microscope images of API X80 surface after 14-day exposures to (**a**) abiotic- and (**b**) sulfate-reducing bacteria (SRB)-inoculated soil suspensions. Pits were observed in both exposures. Pits in the abiotic exposure were isolated. In the presence of SRB, there were clusters of deeper pits. Corrosion products had been removed.
SOURCE: Chen et al. (2021).

increased in the presence of SRB. Additionally, some pits in inoculated samples were connected to form clusters. In the absence of SRB, the maximum pitting depth was 2.32 ± 0.2 μm, while the maximum depth in the inoculated sample was 6.01 ± 0.6 μm. Despite decades of effort to identify diagnostic fingerprints for MIC, most investigators agree that although certain morphologies are consistent with microbially influenced chemistries, MIC cannot be diagnosed solely by the morphology of localized corrosion (Eckert, 2003; Little et al., 2020).

Iron-Reducing Bacteria

IRB can be either strict anaerobes (e.g., Geobacteraceae) or facultative anaerobes that can use oxygen but can also derive energy from the reduction of other electron acceptors (e.g., Fe^{3+}) under anaerobiosis (e.g., *Shewanella* sp.). Fe^{3+} is an efficient electron acceptor, and both anaerobes and facultative anaerobes are able to remove iron oxides from steel in laboratory experiments (Kappler et al., 2021).

Acid-Producing Microorganisms

Specific microbial metabolisms can produce inorganic (e.g., sulfuric) or organic acids (e.g., acetic). The corrosion rate of plain carbon steel in water does not vary with pH values between 4.5 and 9.5 (Coburn, 1978; Uhlig and Revie, 1985), but the activities of acid-producing microorganisms can cause the pH to drop to below 4.5 at the biofilm–metal interface. Under these conditions, protective iron oxide scales dissolve. Gu (2014) concluded that a pH 2 acetic acid solution was much more corrosive than a pH 2 sulfuric acid solution.

Sulfur-oxidizing bacteria (SOB) and archaea can oxidize reduced sulfur species (e.g., sulfides, sulfites, thiosulfates, polythionates, and elemental sulfur) to sulfuric acid. These microorganisms are extremely diverse, both ecologically and taxonomically. Most important, they are multifaceted with respect to the physiology and biochemistry of sulfur oxidation processes, exhibiting different abilities to use specific reduced sulfur compounds as substrates. SOB and SPP are ubiquitous in most soils and are responsible for sulfur cycling. Sulfur cycling has been demonstrated in biofilms and is cited as the mechanism for accelerated low-water corrosion of carbon steel in coastal marine environments.

APB are facultative microorganisms that can produce organic acids through a process of fermentation. Fermentation is a primary means of producing energy by the degradation of organic nutrients anaerobically. Pope (1990) conducted a study of buried steel gas pipelines and concluded that APB were more important to the corrosion than SRB. Gu (2014) reported that APB produced "alarmingly large amounts of organic acids." However, the precise role of APB in MIC is controversial. Pope et al. (1988) and Gu and Galicia (2012) concluded that the acids produced by APB were aggressive to plain carbon steel. In contrast, Mand et al. (2014) reported that the contribution of APB was to provide nutrients for SRB growth. In a field survey, Li et al. (2001) concluded that the maximum corrosion rates were measured when SRB were colocated with APB.

Filamentous fungi, ubiquitous in oxic soils and well known for their ability to convert organic material into organic acids, are not typically cited in the corrosion of buried steel. In general, fungi acidify their microenvironments, including soils, by excreting protons, organic acids, and CO_2 (Gadd, 2010). Soil acidification can contribute to mineral weathering (nutrient cycling) and influence biofilm formation.

Methanogens

Methanogens are archaea capable of producing methane as a metabolic by-product under anoxic conditions. Numerous studies have documented methanogens in corrosive biofilms (e.g., Zhou et al., 2020), and methane formation has been correlated with metal weight loss of steel (Vigneron et al., 2016). However, methane does not react directly with steel. Multiple mechanisms, reviewed by Vigneron et al. (2016), have been proposed for the role of methanogens in corrosion of steel. Proposed mechanisms appear to be characteristic for some lineages but are not widespread among all methanogens and are similar to those suggested for SPP. For example, some methanogens reportedly couple methanogenesis with direct uptake of electrons from Fe^0 (i.e., iron oxidation). Extracellular hydrogenases generated by some methanogens may be involved in the consumption of hydrogen generated by CO_2 corrosion. Corrosion may be related to sulfidogenic dissimilatory sulfur reduction in S^0-rich environments. Additionally, methanogens can contribute indirectly to MIC through interactions with syntrophic APB.

RELATIONSHIP BETWEEN CORROSION AND THE ENVIRONMENT

As described in Chapter 2, there are limited long-term in situ experiments on corrosion of buried steel. Many uncontrolled variables in the most well-known experiments (Logan, 1945; Romanoff, 1957) influenced corrosion rate. The resulting scatter of the data obscured any differences in weathering among the eight alloys tested; that is, the differences in corrosion of the different alloys was not resolvable due to the impact of environmental conditions and the fact that no significant difference could be identified between alloys (Logan, 1945; Romanoff, 1957). Additionally, in a comprehensive review of corrosion of steel in mechanically stabilized earth wall construction, King (1977) acknowledged that all ferrous metal is susceptible to corrosion but that "soil type is the dominant factor in corrosion rather than the type of steel." More recent statistical analysis of the Romanoff (1957) data with linear and multiple regression confirmed that differences in the corrosion of the bare steel and wrought iron (low-carbon) were not identifiable as a function of soil type, stating that "scatter in the measurements resulting from the exposure variables and the natural stochastic nature of underground corrosion overwhelms any differences due to alloy type for this range of alloy compositions" (Ricker, 2010). Despite the fact that the experimental data have a large degree of scatter, these studies indicate that the rate of corrosion is controlled less by the type of steel and more by the physical, chemical, and microbiological properties of the environment in which the steel is buried. Consequently, the variability of the soil and environmental factors, when measured over scales that range from centimeters to meters to kilometers and years to decades, is of primary significance. Given the heterogeneity of the subsurface, the fact that alloy type is a secondary factor in corrosion is not surprising.

It is important to note that the Romanoff (1957) data were based on steel alloys that were available in the early 1900s, and steel design has changed significantly since. It is reasonable to assume that research and development of new alloys could change the types of steels used commonly in practice if steels could be developed that are less susceptible to corrosion, while still being economical at scale. However, because the inherent variability of the parameters of the electrolyte (e.g., soil and moisture) has, to date, obscured any differences between steel types, future studies will benefit from significant focus on the environment, not the alloy. Statistically designed experimental studies with thorough characterization of the soil, moisture conditions, and seasonal variability in climate may allow more sensitive identification of the differences in corrosion rates between different steel alloys, but until those data are available, the type of steel is a secondary consideration in corrosion.

5

Corrosion Protection for Buried Steel

Corrosion of buried steel can be managed through one or some combination of the following four approaches: (1) the service life of the steel can be increased with additional steel incorporated into the steel cross section; (2) the steel can be protected by physical barriers (e.g., coatings and casings); (3) cathodic protection (CP) can be used to modify the electrochemical cell; or (4) the environment itself—the buried soil and surrounding ground—can be modified to reduce the risk of corrosion. As discussed in Chapter 2, the "corrosion allowance" approach employed largely by the geo-civil industries incorporates structural redundancy into design so that if one steel infrastructure component fails through corrosion, the load will be transferred to other infrastructure components to ensure continued performance. Most geo-civil structures have a design life of less than 100 years, so a slow rate of corrosion is tolerated. Although geo-civil design can occasionally include CP (e.g., on bridges), the geo-civil design approach more commonly relies on protective physical barriers and controlled environments in conjunction with increased cross-sectional thickness to maintain a slow corrosion rate. In contrast, oil and gas pipeline industries generally favor a "corrosion protection" approach, which employs protective physical barriers as well as CP to avoid corrosion. Water utilities generally focus little on external corrosion of water pipelines, tolerating leaks and addressing breakages as they happen. Instead, the utilities focus on internal corrosion, water safety, and compliance with the Safe Drinking Water Act (42 U.S.C. § 300f). Given the large spatial distribution of pipe networks, controlling the environment is not economical. This chapter discusses the various means of protecting steel from corrosion and the ways in which each of these systems can fail.

PHYSICAL BARRIERS

Physical barriers such as coatings and casings are a first defense against corrosion. Much steel infrastructure that will be exposed to the subsurface includes a physical barrier. Coatings can range from cement to metal to polymer (see Table 5.1) and may be applied during installation or during repair. The most common physical barriers employed to protect geo-civil buried steel infrastructure are polymeric coatings (i.e., paint or epoxy). Multilayer coatings are often used, which include a primer with good adhesion properties and a topcoat with good corrosion protection properties. Coating systems are chosen based on a variety of factors including environmental and safety regulations, cost, availability, shop or field conditions during application, and effectiveness. Major factors that affect coating performance include the type of exposure and expected service life, surface preparation, adhesion of the coating to the surface, and method of application. It is important to consider that defects can be introduced during application, handling, and installation and that coatings may develop anomalies and mechanical damage as they age.

Oil and gas pipelines generally have considerably thick coatings and can include several types, such as coal tar enamels, fusion-bond epoxy (FBE), two-layer polyethylene (PE), and three-layer PE or polypropylene, among others. Box 5.1 provides a brief description of these different coatings and when they were introduced. Like oil and gas pipelines, water pipeline coatings can include hot-applied coal tar enamel, FBE, or PE, but also commonly include liquid epoxy systems, extruded PE, wax tape, sacrificial metallic coatings, and cement coatings. Some paint systems claim to have extended lifetimes (greater than 50 years) in many underground environments, but many have service lifetimes of only 15–20 years and are permeable to a certain extent to water, oxygen, and ions (Helsel, 2018). Coatings typically lose their efficiency even before the end of their service life.

A coating defect or anomaly poses a threat because it can result in the loss of the physical barrier and allow corrosion. It is possible for all coatings to be defective, be damaged mechanically during installation (e.g., scratched), or suffer degradation in the buried environment from continuous mechanical, chemical, electrochemical, and microbial interactions with the subsurface (Li and Castaneda, 2015, 2017; Li et al., 2016). Defects may be large, such as the pores created by lack of adhesion, or they may be small intrinsic pathways between the spaghetti-like hydrocarbon strands of the polymeric structure of the paint. These defects offer pathways for the transport of water, oxygen, and ions to the steel surface. However, corrosion will only ensue if the coating detaches locally (i.e., disbondment) from the steel. Disbondment allows the formation of several monolayers of water and is the origin of the oxygen differential within the delamination zone, which is required for the corrosion process. It can allow initiation and propagation of the corrosion process, ultimately resulting in the broad failure of the protective coating system. However, if the adhesion of the coating is strong, local disbonding will not readily occur.

Once disbondment is initiated, the coating defect can grow into a blister. This occurs by one of two mechanisms. Anodic undermining or oxide lifting results from the action of voluminous iron oxides. The volume of these corrosion products is greater than that of the metal from which they formed, and that increased volume from reaction can lift the coating, allowing the defect to propagate across the surface. The second mechanism is cathodic delamination in which the oxide corrosion product blocks oxygen, which moves the cathodic oxygen reduction reaction to the blister edge because oxygen and water can be transported through the coating. The reduction of oxygen at the blister edge results in an increase in the pH (see Equation 2.2), which destabilizes the bonds between the polymer coating and the metal surface, promoting an extension of the delaminated region. Cement and metallic coatings are not susceptible to these types of disbonding mechanisms.

Casings (made of PE, concrete, or steel and which are isolated from the steel pipe with a dielectric coating) are physical barriers that are often used on oil pipelines when increased strength or protection is required, such as under rivers, roads, or rail crossings. Casings also have their own set of risks—such as shorted casings where the carrier pipe and casing come in contact with and shield the carrier from CP—but there are installation methods to mitigate this risk, and regular maintenance, corrosion monitoring, and observation of the casings can validate their integrity (Rankin and Al Mahrous, 2005).

TABLE 5.1 Common Types of Protective Coatings

Type of Coating[a]	Example	Comment
Cement coatings	Grout around subsurface anchoring devices	Can lose ability to protect against corrosion after cracking. Difficult to apply at pipe joints. Susceptible to degradation in soils with high sulfate and chloride ion content.
Metallic coatings	Zinc galvanization and aluminum coatings	Once the surface layer is consumed, the protection no longer exists, and the underlying steel will then corrode.
Polymeric	Epoxy coating/paint	Somewhat permeable to water, oxygen, and ions, which often leads to coating failure. The protection decreases with time even before the end of service life.

[a] In some cases, these coatings can be combined.

BOX 5.1
Evolution of Coatings for Buried Steel Pipes

Through the decades, coatings have been modified and improved to protect buried steel infrastructure (Khanna, 2018; Rahman and Ismail, 2013). All of the materials listed below have been applied to buried pipes and are still found in the ground today. Other coatings such as polyvinyl chloride and polyurea are also used for coating buried steel pipes.

1940s–1950s: Coal tar (bituminous enamel), wax, polyethylene (PE) tape, and extruded PE jacket materials were applied in the field.

1960s: Asphalts were introduced. Coal tar coatings, one of the first pipeline coatings to be successfully applied in a plant, were popular until the 1970s.

1970s: Fusion-bond epoxy (FBE) coatings are 100 percent solid, thermosetting powders that bond to steel surfaces as a result of a heat-generated chemical reaction. Formulations consist of epoxy resins, hardeners, pigments, flow-control additives, and stabilizers.

1980s: Polyester and polyether polyurethanes were introduced. Multilayer coatings incorporating FBE were introduced as an alternative to single-layer FBE coatings that were easily scratched during storage and placement. Three-layer polyolefin coatings can consist of an FBE or liquid epoxy primer and copolymer adhesive intermediate or tie layer topcoated with a polyolefin. PE and polypropylene are typical topcoats.

1990s: Liquid epoxies and phenolics were developed as the demand for high-temperature coating applications increased.

2000–Present: Hydrogen permeation barrier coatings have been designed to prevent hydrogen uptake into steel (Korinko et al., 2005). The coatings reduce the possibilities of hydrogen embrittlement, hydrogen-induced cracking, and hydrogen-induced loss of ductility. Corrosion inhibitors—for example, polyaniline and imidazoline—have been added to coatings.

Since metallic coatings (e.g., zinc and aluminum) provide sacrificial CP to the substrate steel, these coatings are discussed in the section on Sacrificial (Galvanic) Cathodic Protection, below.

MICROBIALLY INFLUENCED CORROSION AND COATINGS

Microbially influenced corrosion (MIC) has been identified as a problem for legacy pipeline coatings (i.e., coatings that have been superseded but are still in service) including coal tars, asphalts, greases, tapes, polyvinyl chloride (PVC), PE, and polyester polyurethanes (Little and Wagner, 2002). Soil bacteria, archaea, and fungi can derive nutrients from the water-soluble components of some pipeline coatings. For example, biodegradation of low-molecular-weight components from asphalt coatings by microorganisms results in permeable, embrittled coatings (Little and Lee, 2018). Similarly, plasticizers (e.g., phthalates) can be selectively removed from PVC coatings through water dissolution and biodeterioration. Some polymeric coatings (e.g., polyester polyurethanes) can be directly degraded by extracellular enzymes, acids, or peroxides. Howard (2011) indicated that many soil microorganisms can use polyester polyurethanes as the sole carbon and energy source. Breaches in coatings allow ingress of water and the possibility of biotic or abiotic corrosion.

CATHODIC PROTECTION

CP is an electrochemical technique designed to provide corrosion protection by polarizing a structure in the cathodic direction (i.e., cathodic polarization) using an electrical current. Kuhn (1928) established that polarization to −0.850 V versus a copper–copper sulfate reference electrode (CSE) was required to ensure corrosion protection of a soil-buried cast iron pipe. That criterion, a protection potential, is used today and incorporated into standard practices such as NACE SP0169 (2013) and DNV (2010). The current required to maintain the protection potential (−0.850 V versus CSE) can be provided by sacrificial anodes or impressed current, as described below. The steel structure does not need to be polarized into the immune region of the Pourbaix diagram (see Figure 3.1) for CP to be effective. In fact, polarization to a potential that is too low can result in excessive hydrogen production, which can damage the steel or coatings on the steel. Effective CP need only reduce the corrosion rate of the steel to a value that is approximately 10 times less than that of unprotected steel.

Sacrificial (Galvanic) Cathodic Protection

Sacrificial CP, otherwise known as galvanic CP, uses the differences in the nobilities of two metals to form an electrochemical cell, such that when the two are connected, the less noble metal (anode) will corrode faster than if not connected and the more noble metal (cathode) will corrode slower (see Chapter 4). Buried steel can be cathodically protected with external magnesium or zinc anodes connected (or "bonded") to the steel so that corrosion is primarily displaced from the protected structure to the galvanic anodes. The current flowing through the environment from the sacrificial anode to the structure is accompanied by anodic dissolution of the sacrificial anode material and its irreversible loss. Once the zinc or magnesium anodes are consumed, they need to be replaced for corrosion protection to continue. The service life of the anodes is considered in the design phase of steel structures, but future maintenance and replacement of anodes must continue throughout the life cycle of the steel structures. Sacrificial CP is typically used where the current required for protection is small and the contact resistance between the anodes and soil is limited (i.e., in low-resistivity soil). Remote sacrificial anodes are also used to supplement CP with impressed current in zones of insufficient protection.

Zinc coating (i.e., galvanization) can also provide a type of sacrificial CP for buried steel (see Table 5.1), where the zinc coating is providing protection for the underlying steel. As with bonded remote anodes, when the zinc coating is fully consumed, the protection against corrosion is lost. Several processes can be used to apply galvanization to steel surfaces including hot dipping (ASTM A123/A123M-17, 2017), spin dipping (ASTM A153/A153M-16a, 2016), electroplating (ASTM B633-19, 2019), thermally spraying (AWS C2.23M/C2.23:2018, 2018), or painting (ASTM D6386-16, 2016; ASTM D7396-14(2020), 2020). The coating thickness and the adherence of the galvanization to the steel vary depending on the technique used to coat the steel. In general, hot-dip galvanization results in a thicker coating and a metallic bond with the steel. Steel that includes threaded parts or other pieces that fit together is galvanized through spin dipping or electroplating. The thermally sprayed process is sometimes used for steel piles while painting is only used as a repair for nicks that may be observed prior to installation of buried galvanized steel. Galvanizing is common for less-critical pipelines (e.g., culvert piping for which the consequence of failure is not catastrophic; see Box 5.2), due to the limited amount of zinc typically applied to the steel (e.g., 3- to 5-mil zinc coating thickness). Galvanization is not used for critical pipelines such as oil, gas, or transmission water pipelines (with high consequence of failure), which rely on impressed current CP (ICCP). Installation of external anodes is more practical than galvanizing since these anodes provide a larger amount of sacrificial zinc or magnesium, can be monitored for their efficiencies, and are replaceable once consumed.

Aluminum coating on steel (referred to as aluminized steel) is commonly used for culvert pipe application. The aluminization process results in a dual coating with an inner intermetallic brittle layer approximately 15 µm thick (composed of Fe_2Al_5) and a nearly pure outer soft aluminum matrix layer approximately 30 µm thick (Caseres and Sagues, 2005). Aluminized pipes are commonly ribbed or corrugated for structural strength (corrugation process to occur after aluminized coating is applied to the steel). Ribbed pipes have better hydraulic efficiency and are often preferred.

BOX 5.2
Culvert Protection and Performance

Culverts are unique compared to other buried steel structures because water is deliberately directed toward a culvert whereas in other applications, surface and subsurface drainage is such that water is directed away from the structure (and often toward a culvert). Culverts may undergo corrosion on the inside surfaces that are exposed to flowing water that carries sediments and debris, and on the outside surfaces that are exposed to soil and the subsurface environment. Culverts are often protected from corrosion with galvanized or aluminized coatings. Although aluminum is harder than zinc, resulting in better resistance to abrasion, aluminum may not remain passivated in environments where the pH is less than 4 or greater than 8.5. In these environments, aluminum coatings will undergo higher rates of corrosion than galvanized coatings.

For example, Akhoondan and Sagüés (2013) studied the conditions contributing to observations of high corrosion rates for installations of a specific type of aluminized steel drainage pipes at inland locations within Florida where aluminized pipe was exposed to a high-pH environment. The aluminization was applied to base steel via the hot-dip process, resulting in an ~50-µm-thick coating with outer and inner layers of commercially pure Al and alloys of Al-Fe, respectively (ASTM A929/A929M-18, 2021). Pipe installations were in Florida and were backfilled with limestone aggregates. Water in contact with the backfill was expected to equilibrate with CO_2 in the atmosphere and to develop a near-neutral pH, which is a favorable condition for the performance and durability of Type 2 aluminized steel (AST2). Under these conditions, a passive film was expected to form and protect the Type 2 aluminum coating from corrosion, but in many cases, unexpected early corrosion of AST2 pipes was observed. Akhoondan and Sagüés (2013) performed laboratory experiments where they simulated service conditions by allowing water to flow slowly through the aggregate while they observed changes in pH and conductivity. They observed that the pH of the pore water was steadily above 9, which is an unfavorable condition for the performance of aluminum. Akhoondan and Sagüés hypothesized that the flow conditions in Florida did not render enough time for water to reach chemical equilibrium with CO_2 from the atmosphere. Thus, the pH was not buffered, and a stable passive film layer could not form to protect the surface of the AST2.

Another study by Akhoondan et al. (2008) also indicated extensive corrosion near coating breaks (due to gross manufacturing defect) in a near-neutral regular soil environment where passive film is stable. This was in agreement with premature failures of aluminized culvert piping in Florida.

While the zinc coating (on galvanized steel application) is subject to continuous corrosion to provide protection, in the case of aluminized steel, a thermodynamically stable thin passive film of aluminum oxide is expected to form rapidly and prevent further corrosion. While the presence of passive film can increase the durability of aluminized steel by several folds compared to galvanized steel (Cerlanek and Powers, 1993), it may also cause localized corrosion and premature failure in areas of coating defects and blemishes. Coating blemishes may occur due to improper storage, handling, installation, or manufacturing defects in corrugation process. Additionally, aluminized passive film is only stable in the pH range of 4 to 8.5. In the pH ranges below 4 and beyond 8.5, the passive film will dissolve and no longer provide protection, and the corrosion rate of aluminized steel will be higher than that of galvanized steel.

Impressed Current Cathodic Protection

In some applications, the potential difference between sacrificial anode(s) and the steel structure cannot generate sufficient current for protection to occur. In other situations, repeated replacement of consumable anodes is not possible. To overcome the limitations inherent in the use of galvanic CP, direct current (DC) from any power source can be used for ICCP systems (Beavers, 2001). Typically, alternating current (AC) power sources are used in conjunction with rectifiers to convert AC to low-voltage DC (see Figure 5.1). The anodes for ICCP are made from nonconsumable materials such as chromium-bearing high-silicon cast irons or conductive oxides. ICCP is used extensively in oil, gas, and large water transmission pipelines. The advantage of ICCP is that the anode does not need to be replaced, thus reducing maintenance costs. However, installation costs are higher due to the cost of required hardware and electrical work. Additionally, aboveground ICCP hardware must be safe and secure. ICCP also requires an AC power supply over the life of the CP system, which adds to maintenance costs. Finally, ICCP has a greater likelihood of creating stray-current interferences on other steel infrastructure in the vicinity.

Cathodic polarization of a buried steel surface immediately reduces the kinetically controlled iron dissolution rate and increases the cathodic reduction rate on the steel surface. Two cathodic reactions at the metal surface are responsible for a progressive change in the corrosivity of the area around a cathodically protected surface: reduction of oxygen and reduction of water (see Equations 2.2 and 2.4). Both reactions produce an accumulation of OH^- at the surface, contributing to increased local pH (Leeds and Leeds, 2015). The magnitude of the pH change depends on the extent of cathodic polarization and the local environmental conditions, including soil chemistry and microstructure, hydrodynamics (stagnant or moving water), and the presence and activities of microorganisms (Angst, 2019). The alkaline zones generated by cathodic polarization can extend from the steel surface into the soil for centimeters to decimeters and can persist for hours after polarization ceases (Angst, 2019; Brenna et al., 2017). Alteration of the local chemical environment is an essential component for corrosion control of buried steel by CP because steel corrodes slower in high pH environments even in the absence of CP (Angst, 2019).

In both galvanic CP and ICCP, the anodic reaction is shifted primarily from the protected metal structure to the remote anode. Both are used in combination with coatings to reduce the spatial area requiring CP. In such applications, the role of CP is primarily the protection of exposed metal where coatings have been damaged or degraded. The need and design for CP are determined through field and laboratory tests (see Chapter 6). Certain geotechnical conditions (e.g., the presence of coarse-grained soils) limit the installation of CP due to both high resistivity and the difficulty of anode trenching.

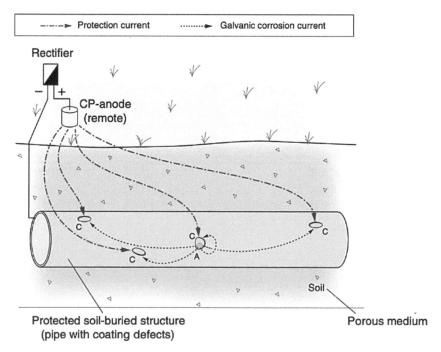

FIGURE 5.1 Alternating current is converted to direct current using a rectifier to produce impressed current cathodic protection (CP). In this example, the CP-protected buried pipe is experiencing galvanic corrosion between anodic (A) and cathodic (C) sites on the structure.
SOURCE: Modified from Angst (2019).

Assessment of Cathodic Protection

Practical application of CP for buried steel is complex. Both types of CP require electrical and ionic connection between the anodes and all surfaces to be protected. Ionic connectivity means that ionic current can flow from the anodes through the soil electrolyte to reach the steel surface. Spatial variations in the underground environment can limit transport and cause local variations in moisture near the steel structure that disrupt the overall path of current to the anode. Environmental pH, temperature, oxygen content, ionic concentrations, biological activity, and resistivity influence the current in the electrochemical system and ionic transport at the steel–subsurface interface. Temperature can accelerate chemical and electrochemical reactions, increasing the current needed for protection. Fundamental mechanisms for CP effectiveness are somewhat controversial. For example, it is unclear whether CP results in a reduction in the number and size of corroding sites or a reduction in the corrosion rate of those sites (Angst, 2019).

Another matter of debate involves the relationship between cathodically polarized steel and soil microorganisms. For example, Miyanaga et al. (2007) suggested that bacterial cells within an artificial preexisting biofilm were killed or damaged by the pH increase during CP. However, several researchers have demonstrated that CP cannot prevent biofilm formation and may attract microorganisms (Guan et al., 2016; Jansen et al., 2017). Specific influences are difficult to tease out because there are often differences in electrolytes, microorganisms, and experimental conditions. The results of polarization experiments conducted in liquid media cannot be interpreted as significant to cathodically polarized steel in soil.

Much of the confusion related to microorganisms and the effectiveness of CP is specific for sulfate-reducing bacteria (SRB). In anaerobic environments, SRB generate hydrogen sulfide (H_2S) that dissociates to H^+, bisulfide (HS^-), and sulfide (S^{2-}). Most CP standards recommend protection potentials that are at least 0.100 V more negative than the normal reference condition (i.e., -0.950 V_{CSE}) if there are insoluble ferrous sulfides (NACE SP0169, 2013; DNV, 2010). Barlo and Berry (1984) concluded that even more negative potentials were required to prevent corrosion of steel when SRB were active, but Jansen et al. (2017) demonstrated that once biofilms were established, it was difficult or impossible to prevent MIC irrespective of applied potential. Furthermore, increased pH that typically accompanies cathodic polarization can be significantly reduced by the presence of biofilms that provide a chemical pH buffering capacity. Overall, there is general agreement that more negative protection potentials are needed for CP in the presence of iron sulfides, but there are unwanted consequences of extreme cathodic polarization. Electrical fields and high pH generated by cathodic polarization can cause coating disbondment (Pope and Morris, 1995). The additional negative potential applied to protect against MIC also increases the generation of atomic hydrogen, which can enter and degrade the steel (Kim, 2002).

CONTROL OF ENVIRONMENT

Site conditions, location, climate, and properties of the fill, native soils, or rock are factors that can affect corrosion. The climate, and often the location, cannot be altered, but the site and ground conditions may be controlled. Mitigation strategies often use lime to increase the pH, compounds that inhibit corrosion, or methods to control the amounts of dissolved solids by removing soluble salts. These mitigation strategies are described in more detail in Table 5.2.

TABLE 5.2 Mitigation Strategies to Control Corrosivity of the Environment

Method	Comment
Mix lime with fill sources prior to or during placement	Results in a desirable, homogeneous distribution of the lime or corrosion inhibitor.
Inject lime into fills during service	Spacing and dosage per injection site need to be properly selected such that the compounds permeate uniformly throughout the fill.
Install barriers that intercept and treat groundwater	Barriers may include chimney drains or blankets (Berg et al., 2009). These barriers can protect fills from contaminants (e.g., deicing salts applied to pavements, phosphates from applications of fertilizers).
Leach and drain soils to remove salts	Remove soluble salts by continuously spraying the surface with clean water until the leachate no longer includes impurities. This approach requires that an internal drainage system is incorporated within the fill to effectively collect and remove leachate from the system.
Separate soils by grain size (i.e., scalping) to remove fractions containing soluble salts	Finer fractions (i.e., those passing the #40 sieve) may be more corrosive due to lower pH and resistivity and higher sulfate concentration.

SOURCES: Thapalia et al. (2011); Timmerman (1992).

6

Standard and Evolving Subsurface Characterization

Prior chapters introduced types of steel, the earth materials in which steel is placed, and the corrosion processes that affect underground steel infrastructure. Chapter 4 established that the propensity for and rate of steel corrosion are more greatly influenced by the characteristics of the subsurface environment rather than the composition of any of the steels that are currently economical, practical, and in common use for buried-steel applications. Therefore, a comprehensive understanding of the subsurface is essential to model accurately the corrosion of buried steel. From a corrosion risk perspective, subsurface characterization includes identifying components of the subsurface that may contribute to the corrosivity of the environment. Many test methods used in corrosion investigations were adapted from methods developed for application in the field of agronomy. While tests completed in the laboratory setting yield precise measurements at a single point in space and time, the results may not necessarily represent field conditions because of changes in the biogeochemistry during transport and sample preparation. These tests are referred to in this report as "screening tests." In contrast, in situ field analyses are typically completed with field parameters relatively intact but often are limited in their resolution given the heterogeneity of the subsurface. Given the difficulties characterizing a spatially and temporally variable subsurface, characterization techniques— both laboratory and field—at different scales are often combined. These analyses are used to assess the magnitude of subsurface corrosivity, estimate corrosion rates, and determine appropriate corrosion protection measures. However, corrosion rate estimates often do not include propagated error, nor do they reflect possible temporal variability in conditions, nor on how temporal variability will be affected by global climate change (e.g., changing temperatures, rainfall and soil-moisture conditions, changing soil-ionic content as a result of sea level rise [see, e.g., Zhang et al., 2022]). Additionally, the lack of field data from long-term and well-controlled experiments (see Chapter 2) poses a significant challenge for the assessment of the efficacy and uncertainty of predictive models that integrate many of these subsurface characteristics (see Chapter 8). The committee finds it is unable to draw conclusions regarding the efficacy of current practices.

This chapter explores current field and laboratory subsurface characterization techniques that inform the prediction, assessment, and modeling of soil corrosivity and corrosion rates of steel structures in the subsurface, as well as how those techniques are evolving. Although there are numerous techniques used to characterize the subsurface, this chapter focuses on those associated with understanding corrosivity of the subsurface environments and the corrosion rates of steel placed in them. The chapter focuses on understanding corrosivity and corrosion rates of steel buried in soils because steel is less commonly in direct contact with in situ rock (steel piles may be driven through soil to sit on top of in situ rock, and mechanical reinforcing rock bolts may be rarely or temporarily inserted into rock; see Table 2.1).

The chapter first describes the subsurface characterization that occurs before and during installation, then discusses subsurface characterization that occurs during operation and after failure in a forensic setting, and finally describes some emerging characterization methods. Note that there are often several methods that seem to be similar or seem to provide similar results. This is a result of different methodologies being developed for different industries or by different equipment manufacturers who have developed their own patents. The existence of these multiple methodologies that target similar parameters may be interpreted to signify that no single methodology has yet been developed that provides all the information needed, and that each of the methods has distinct advantages and disadvantages. The committee does not attempt to rank or recommend the use of one methodology over another but describes these methods with the understanding that there are uncertainties inherent in the use of any of them. Box 6.1 provides a list of common terms used in this chapter and the remainder of this report.

STANDARD PRACTICE BEFORE AND DURING INSTALLATION

The evaluation of soil corrosivity often depends on standards of practice in a local geographic region. If a given locality has not been considered historically to be corrosive, then there may be only modest in situ or laboratory testing to determine corrosivity. On the other hand, if a given locality is known to have properties conducive to corrosion, such as high concentrations of chloride, or if a locality has not previously been extensively studied, then explicit in situ and laboratory testing from samples may be performed. Most soil corrosivity testing is conducted traditionally through laboratory testing of bulk samples collected by disturbed sampling at the site and not from in situ field testing.

**BOX 6.1
Definition of Terms**

Characteristic: Observation of a material that does not vary regardless of testing condition (e.g., grain size). However, estimates of the tested characteristic may change based on assumptions of a given test method or the uncertainties inherent to the test.

Property: Measurable quantity that changes based on testing conditions that are then used in design (e.g., pH of diluted sample, resistivity, stiffness). Often depends on soil characteristics.

Parameter: A variable (e.g., temperature, stress state, fluid saturation, mineralogy, porosity) that affects a property.

Field test: Test performed in situ to measure engineering properties or characteristics of soil with field parameters relatively intact (e.g., strength, stiffness, or hydraulic conductivity).

Laboratory test: Test performed in a laboratory to measure engineering properties or characteristics of soil.

Screening test: Test performed according to a standard procedure in either laboratory or field setting to classify soil or groundwater properties. Field parameters such as stress state or fluid chemistry are likely not preserved (e.g., pH of diluted sample).

Index test: Test that produces a result used to compare different soils or materials in terms of anticipated behaviors or performance, but that does not quantify material property or characteristic directly (e.g., Atterberg limits—liquid limit, plastic limit, and plasticity index).

At the start of any investigation, whether for gathering general geotechnical property information or specifically to characterize corrosivity of a site, the geotechnical engineer will search for historical information from, for example, existing geological maps from the U.S. Geological Survey (USGS) or other sources, and from boring and monitoring well logs proximate to the site that may have been cataloged by local building authorities or other project owners. Local and city building authorities may have records (often publicly available online). Such historical information can guide the need for additional study. Good practice for geotechnical engineers is to also consider the possible presence of stray currents (see Box 6.2).

If available information is not sufficient to inform decision making related to design, corrosion control, or operation and maintenance, then the engineer needs to develop a field and laboratory investigation to study soil corrosivity. Selected laboratory testing to characterize corrosivity will be conducted on bulk soil samples collected during the geotechnical investigation to establish design parameters. The amount and type of corrosivity testing typically is based on a balance of anticipated variation in overall site conditions and costs. In practice, it is rare to conduct laboratory corrosivity testing on bulk samples from every boring location and at every depth as might typically be done to characterize geotechnical and physical properties for design. As a result, heterogeneity in those properties that might be conducive to corrosivity may not be captured.

In some cases, especially for large geo-civil infrastructure projects or oil and gas pipelines, in situ field corrosivity testing may also be conducted to complement laboratory testing of field samples. The principal reason is that steel may penetrate multiple soil strata, and the lateral character of the soil conditions may also vary significantly over the project site. The more common field methods currently in use to assess corrosivity include electromagnetic measurements and electrical resistivity. These methods are discussed later in this chapter.

Laboratory Methods

Laboratory methods provide quantitative measurements of soil and groundwater characteristics and properties under controlled and repeatable testing conditions. These tests are designed to measure a single variable (e.g., pH) for a sample taken at a single point in time and in space; consequently, the results are inherently limited in their spatial and temporal scale. Laboratory methods can be divided into laboratory tests, screening tests, or index tests (which do not quantify material properties or characteristics directly) (see Box 6.1). Procedures for measuring soil electrochemical properties are described in a variety of standards and recommendations from the American Association of State Highway and Transportation Officials (AASHTO), ASTM International, the American Water Works Association (AWWA), the American Public Health Association, the Soil Science Society of America, the USGS, the U.S. Environmental Protection Agency, and the Natural Resources Conservation Soil Survey. Different standards for laboratory methods serve specific industries and different locales, and the standards often apply the same general techniques but may differ in recommendations for sample preparations or treatments. Some state departments of transportation have developed their own testing procedures (e.g., California, Nevada, North Carolina, Pennsylvania, and Texas) because of unique construction practices, such as the use of coarser materials for fill due to local availability. The choice of applicable standards is based on the infrastructure type and location.

BOX 6.2
Stray Currents

Characterization of site corrosivity needs to include any possible sources of stray currents. Historically, stray currents and associated corrosion have been controlled through cooperative efforts of transit agencies and affected parties. Various engineering approaches have been implemented aimed at controlling the generation of stray currents or controlling the way in which stray currents flow. However, the state of the art of stray-current control is substantially the same today as it was in the early 1900s (McCarthy et al., 1988). Stray-current control guidelines are described in Transit Cooperative Research Program Research Report 212 (Flounders and Memon, 2020).

For example, practices related to a transportation asset such as a mechanically stabilized earth (MSE) wall will generally follow AASHTO standards unless the MSE wall is in a jurisdiction with its own standard. Agencies that build infrastructure without their own governing standards (e.g., the U.S. Bureau of Reclamation) often use international guidelines such as those written by ASTM International.

Many tests performed to identify soils that contribute to corrosion are consensus standard tests, which balance factors such as test complexity and analytical precision with cost and limited accessibility to sophisticated laboratory equipment. In Tables 6.1 and 6.2, both ASTM and AASHTO standards are cited when they are equivalent (e.g., AASHTO T 265-15, 2019 and ASTM D2216-19, 2019 for moisture content). In other cases, more than one test standard exists for measurement of a given property, but there are differences in terms of sample treatments and preparations. For example, resistivity and pH can be measured via AASHTO T 288-12 (2016) and AASHTO

TABLE 6.1 Example Laboratory Test Methods Consistently Cited as Important in the Identification of Corrosive Soils

Characteristic/ Property	Importance	Method	Example Standards
Soil classification	Unified Soil Classification System used to categorize soils into designations: gravel, sand, silt, or clay.	Determine soil grain size using sieve analysis or hydrometer and measure plasticity of fine-grained soils.	ASTM D422-63 (2016) ASTM D4219-08 (2017) ASTM D4318-17e1 (2018) ASTM D2487-17e1 (2020)
Resistivity (inverse of conductivity)	Lower resistivity values generally correspond with increased corrosivity. This is a screening test and a worst case for corrosivity because samples are tested when the soil is saturated. Although the lowest resistivity may be when the soil is saturated, it is not the most severe condition for corrosion, which is 60–85 percent saturated.	Measured with standard resistivity meter, either alternating current or 12 V direct current. Approximately 1,500 g of soil are confined between two stainless steel electrodes. Water content is incrementally increased with distilled water until a minimum resistivity value is determined.	AASHTO T 288-12 (2016) ASTM G187-18 (2018) ASTM WK24621 (2015) ASTM G57-20 (2020)
Moisture content (i.e., water content)	Corrosion reactions will not occur in dry soils, and increase with moisture content. Data from this test indicate field conditions.	Gravimetric ratio of mass water and mass solids (soil or rock).	ASTM D2216-19 (2019) AASHTO T 265-15 (2019)
pH	Rates of general corrosion and pitting increase as pH decreases and liberates [H^+]. This is a screening test because the test is performed on a diluted sample.	Measured for soil in slurry form after equilibrating with distilled water for >1 hr. Measurement taken using a pH electrode immersed in the slurry and reported at 25°C, or measured on dry soil using pH paper or pH meter.	AASHTO T 289-91 (2018) ASTM D4972-19 (2019)
Chloride/sulfate/ soluble salt content	Quantifies mass of water-soluble salts in soils, which are commonly sulfates or chlorides. Increasing salt content decreases soil resistivity, which is correlated with increased corrosivity. Data from this test indicate field conditions, because it is performed on pore water extracted from the sample. Sulfates can also promote the growth of sulfate-reducing bacteria and microbially influenced corrosion.	Soil pore water is separated from the solids and chloride/sulfate/soluble salt content is determined with a refractometer, titration, ion exchange chromatography, or ion selective electrodes.	Chloride: AASHTO T 291-94 (2018) Sulfate: AASHTO T 290-95 (2020) ASTM C1580-20 (2021) Sulfate and chloride: Tex-620-J (2002) Tex-620-M (2018) Soluble salts by refractometer: ASTM D4542-15 (2016) Multiple anions by ion chromatography: ASTM D4327-17 (2019)

TABLE 6.2 Example Laboratory Test Methods for Identifying Corrosive Soils

Characteristic or Property	Importance	Method	Example Standards
Sulfide	Sulfides may indicate the presence of sulfate-reducing bacteria and microbially influenced corrosion. Data from this test indicate field conditions.	Uses ion-selective electrode to measure sulfide concentrations in aqueous solutions.	ASTM D4658-15 (2017)
Dissolved oxygen	Increases in dissolved oxygen increase corrosivity due to biological activity and geochemical reactions. Data from this test indicate field conditions.	For most groundwater, measured using a standard electrochemical dissolved-oxygen probe that can be either amperometric or potentiometric systems.	ASTM D888-18 (2018)
Organic content	Increased levels of organic matter decrease pH, which increases corrosivity. Organic matter can also participate in redox reactions Finally, organics can promote microbially influenced corrosion (Vazquez, 2014). Data from this test indicate field conditions.	Measurement of change in sample mass during combustion in a muffle furnace (loss on ignition). Percent organic matter is reported as the mass loss as the sample is combusted, which overestimates organic content because there are other sources of mass loss (e.g., breakdown of carbonates or dehydration of water-bearing minerals).	AASHTO T 267-86 (2018)
Alkalinity (bicarbonates)	Measure of the acid-neutralizing capacity of a water (Alkalinity = $[HCO_3^-]+2[CO_3^{2-}][OH^-]-[H^+]$). This test is a screening test when performed on a diluted sample and a field test when performed on a sample of pore water.	For freshwater and saltwater (or brackish water), measured using an acid titration curve.	ASTM D1067-16 (2016) ASTM D3875-15 (2017)

T 289-91 (2018), or via ASTM G187-18 (2018) and ASTM G51-18 (2021). Other examples include variations in the methods used to test salt content using pore waters (ASTM D4542-15, 2016), or extracts from the surfaces of air-dried samples (AASHTO T 290-95, 2020; AASHTO T 291-94, 2018).

Because of the complexity of the soil electrochemical system, comprehensive identification and quantification of the soil properties that influence corrosion are challenging. Properties such as resistivity, water content, soil saturation, and pH are interdependent and vary spatially and seasonally. As a result, many different classification schemes have been developed to evaluate the aggressiveness of an environment based on single or multivariable assessments of the electrochemical property. For example, AASHTO (2002) outlines minimum or maximum electrochemical properties and defines environments that meet these conditions as corrosive (e.g., pH less than 5 and greater than 10; resistivity of less than 3,000 ohm-cm; chlorides greater than 100 parts per million; and sulfates greater than 200 parts per million). In contrast, buried galvanized steel structures (e.g., culverts) are assessed in Great Britain using a multivariate classification system that considers particle size, plasticity, resistivity, pH, sulfates, chlorides, sulfides, and the presence of the groundwater table and drainage conditions at the site (Brady and McMahon, 1994). Other multivariate soil corrosivity assessment schemes include one developed by the Florida Department of Transportation (FDOT, 2018), which also uses minima or maxima (see Table 6.3). In contrast, the criteria proposed by the AWWA (see Table 6.4) are based on a five-property cumulative scoring system, in which soils that have fewer than 10 points are considered noncorrosive to steel, and those with 10 or more points are considered aggressively corrosive (ANSI/AWWA C105/A21.5, 2018). Analysis of 12 different classification and rating schemes (see Table 6.5) indicates that there are roughly 20 soil characteristics and properties that have been identified as contributors to corrosion of buried steel. Of those, the following seven are consistently cited as important in the identification of corrosive soils: soil type, groundwater table, moisture content, resistivity, pH, chlorides, and sulfates (NCHRP, 2017). The laboratory measurement of resistivity, which is cited as an important means of screening for corrosivity in 10 of the 12 surveyed classification schemes, is discussed in more detail in the following section.

TABLE 6.3 Criteria for Determining If the Substructure Is Moderately, Slightly, or Extremely Corrosive

			Steel
Classification	Environmental Property	Unit	Water/Soil
Extremely aggressive (If any of these conditions exist)	pH		<6.0
	Cl	ppm	>2,000
	Resistivity	ohm-cm	<1,000
Slightly aggressive (If all of these conditions exist)	pH		>7.0
	Cl	ppm	<500
	Resistivity	ohm-cm	>5,000
Moderately aggressive	This classification must be used at all sites not meeting requirements for either slightly aggressive or extremely aggressive environments.		

NOTE: Cl = chloride content; pH = acidity (−log of H+ ion concentration, −$\log_{10}$H+; potential of hydrogen).
SOURCE: FDOT (2018).

TABLE 6.4 American Water Works Association Point System for Evaluating Soil Corrosivity[a]

Soil Characteristic or Property	Values	Points
Resistivity (ohm-cm)	<1,500	10
	≥1,500–1,800	8
	>1,800–2,100	5
	>2,100–2,500	2
	>2,500–3,000	1
	>3,000	0
pH	0–2	5
	2–4	3
	4–6.5	0
	6.5–7.5	0
	7.5–8.5	0
	>8.5	3
Redox potential (mV)	>100	0
	50–100	3.5
	0–50	4
	<0	5
Sulfides[b]	Positive	3.5
	Trace	2
	Negative	0
Moisture	Poor drainage (continuous wet)	2
	Fair drainage (generally moist)	1
	Good drainage (generally dry)	0

[a] Soils with less than 10 cumulative points are considered noncorrosive to steel, and those with 10 or more points are considered corrosive or aggressive.
[b] If sulfides are present and low or negative redox-potential results are obtained, give 3 points for this range.
SOURCE: ANSI/AWWA C105/A21.5 (2018). Reprinted with permission from American Water Works Association. Copyright © 2018. All rights reserved.

TABLE 6.5 Classification and Rating Schemes for Identifying Corrosive Soils: Contributing Properties[a]

	DVGW GW 9 (2011)[b]	DMRB BD 42/00 (2000)[c]	DIN 50929-3 (1985)[d]	CEN 12501-2 (2003)[e]	AASHTO (1990)	Brady and Mc-Mahon (1994)[f]	FDOT (2018)[g]	Bureau of Reclamation (NRC, 2009)[h]	ANSI/AWWA C105/A21.5 (2018)[i]	Lazarte et al., (2003)[j]	CEN 14490 (2010)[k]	AGA (2010)[l]
Soil type	X	X	X	X	X	X					X	X
Soil condition (disturbed/undisturbed)			X		X							
Grain size						X						
Plasticity Index						X						
Vertical/horizontal homogeneity			X	X	X							
Water table level	X	X	X	X	X	X			X			
Resistivity	X	X	X		X	X	X	X	X	X	X	
Moisture content	X	X	X		X		X		X		X	X
pH	X	X	X		X	X	X		X	X	X	X
Carbonate	X											
Total alkalinity/acidity (buffer capacity)			X									
Chloride	X	X	X		X	X	X			X		X
Sulfate	X	X	X		X	X	X			X		
Sulfides/hydrogen sulfide		X	X			X			X			
Cinder and coke	X				X	X					X	
Redox potential					X				X			
Sulfate-reducing bacteria					X							
Contamination by deicing salts, manure, fertilizers, leaking sewer, industrial pollution				X							X	
Stray currents			X	X						X		

[a] Rows in gray are those that are commonly cited as important across classification and rating schemes.
[b] German Gas and Water Works Engineers' Association Standard.
[c] Modified Eyre and Lewis System.
[d] German updated DVGW procedure.
[e] European Standard.
[f] UK Corrosivity Classification Test for Soil.
[g] Florida Department of Transportation.
[h] Review of the Bureau of Reclamation's Corrosion Prevention Standards for Ductile Iron Pipe.
[i] Installation procedures for polyethylene encasement to be applied to underground installations of ductile iron pipe.
[j] Ground Corrosion Potential.
[k] European Criteria for Assessing Ground Corrosion Potential after Clouterre (1993).
[l] American Galvanizers Association.
SOURCE: Data from NCHRP (2017).

Laboratory Electrical Resistivity Testing

When the subsurface is exposed to an induced or natural electrical field, the response can be described as the amount of stored electrical energy (permittivity) or the amount of lost electrical energy (conductivity) (Knight and Endres, 2005). Permittivity and conductivity measurements provide the same information, but it is conductivity and its inverse—resistivity—that are most associated, in common practice, with soil corrosivity. Because corrosion is a complex function of the properties of the buried steel, the soil electrolyte, dissolved anions and cations, and the microbiological environment, the resistivity of the soil is not a measure of soil corrosivity (see Box 6.3; Chapter 2). However, when combined with measurements of additional soil parameters, resistivity can be a useful input for corrosion investigations. Consequently, it is a frequently measured parameter in both field and laboratory investigations.

Soil electrical conductivity is the temperature-dependent ability of soil to conduct electric current through the volumes of the solids, liquids, and gases that comprise soil. Fluctuating temperatures will result in different soil electrical conductivity measurements if all other properties remain the same. As a result, soil electrical conductivity measurements often need to be normalized to a reference temperature, particularly in a laboratory setting. However, bulk electrical conductivity is also strongly dependent on total salt content, soil mineralogy, porosity, pore geometry, and degree of saturation (Santamarina et al., 2005).

Laboratory methods to determine resistivity (the inverse of conductivity and a commonly used indicator of soil corrosivity) fall into two general categories: (1) measurements of voltage drop in response to a current applied to a soil box sample, and (2) conductivity measurements on aqueous solutions that include soluble salts extracted from samples of earthen materials (leachates). In the first category, resistivity (ohm-centimeters) of soil samples in a soil box (dimensions of length [L] and cross-sectional area [A]) is measured by inducing a known electric current (I) between two electrodes placed in the soil sample and measuring the resulting voltage potential difference between the electrodes (ΔV). The resistivity can then be directly calculated as $\rho = \frac{\Delta VA}{IL}$ (where ρ is resistivity). For the second category, conductivity is measured on an aqueous solution extracted from soil, or on soil–water mixtures at varying moisture contents (e.g., AASHTO T 288-12, 2016; ASTM G187-18, 2018). Sample preparation treatments such as air drying, sieving, mixing, settling and curing times, and filtering may vary among tests. The state of practice is to assess the resistivity in the saturated state and have the minimum resistivity represent the "worst-case" corrosivity of a particular soil. Minimum resistivity is a main AASHTO criterion to characterize a

---

BOX 6.3
Resistivity of the Soil Is Not a Measure of Soil Corrosivity

Soil resistivity has historically been used as a key, single parameter to indicate soil corrosivity (e.g., King, 1977; Romanoff, 1957). There are parallels between the electrochemical process of corrosion of steel buried in soil (where an electrolyte conducts ionic current between anodic and cathodic sites) and soil electrical resistivity (primarily a measurement of ionic transport in an electrolyte). However, soils are complex, heterogeneous materials and their electrical resistivity is a temperature-dependent response that is a function of the relative volume of the three phases, the mineralogy of the solid phase, the arrangement and interconnectivity of the soil particles, and the pore fluid composition (Santamarina et al., 2005). This wide range of controlling factors also leads to somewhat ambiguous results in the absence of supporting information for all applications of electrical resistivity (Binley and Slater, 2020), including determining soil corrosivity. Therefore, two soils of the same electrical resistivity may exhibit widely varying soil corrosivity. One thing is certain: electrical resistivity has never been successful at accurately determining a corrosion rate. Because of the strong parallels between the dynamically related factors that are believed to influence soil corrosivity and electrical resistivity, resistivity will likely remain a useful input in soil corrosivity studies and therefore has a significant presence throughout this report. However, other parameters discussed in Chapters 2 and 5 should also be considered for estimation of corrosion rates.

soil as aggressive or nonaggressive.[1] The two-point resistivity measurements usually employed in these standards are sensitive not only to the resistivity of the soil sample but also to the contact resistance between the electrodes and the sample. While the contribution may be minimal in the case of conductive (and hence more corrosive) soils, it can be considerable for highly resistive soils.

Field Geophysical Methods

Field geophysical methods are commonly used as part of a geotechnical site investigation plan. By measuring subsurface geophysical properties—which are controlled by physical, chemical, and microbial characteristics of the subsurface—geophysicists can identify changes in stratigraphy, the location of the groundwater table, buried artifacts (e.g., infrastructure), and other subsurface characteristics. Although these measurements are volume averaged, they are still highly valuable. Field geophysics can be used before sampling to optimize a geotechnical site investigation plan (i.e., to identify specific sampling locations) or to map between sampling locations and reduce design and construction risks. Measurements are taken in situ, either through surveys conducted above or at the ground surface or in a borehole (downhole). Common geophysical methods include the electrical resistivity, electromagnetic induction (EMI), seismic reflection and refraction, seismic surface-wave analysis, and ground-penetrating radar. Each method has advantages and disadvantages in its ability to characterize a site and should be chosen based on the objective of the site characterization plan and the general geology of a site. Van Nostrand and Cook (1966) note that subsurface electrical conductivity measurements vary more than any other physical soil property, making electrical conductivity measurements powerful indicators of changes in subsurface. Electrical resistivity and EMI are most commonly used as field methods when considering soil corrosivity, likely because both measure electrical resistivity and because there are laboratory techniques that can be conducted in parallel.

Because the subsurface is highly heterogeneous, field electrical resistivity methods require different considerations and may produce different results from those applied in the laboratory. Field resistivity measurements are likely to be affected not only by soil mineralogy, porosity, pore geometry, degree of saturation, and salinity but also by biogeochemical variability in the subsurface (e.g., as a result of increased water retention by biofilms in sandy soils). Volk et al. (2016) demonstrated that reduced hydraulic conductivity created by a biofilm increases water holding capacity, which, in turn, increases the degree of saturation and therefore the electrical conductivity. Anthropogenic processes can also alter the electric response of the subsurface. Grouts may reduce pore space and alter the conductivity, and anthropogenic dewatering or shifting of the groundwater table will alter the electrical response as a result of changes in the degree of saturation and ionic concentration of the pore water. And finally, physical infrastructure such as power transmission lines, light transit systems, or impressed current cathodic protection systems may produce stray currents (see Chapter 2). Stray currents will affect resistivity measurements taken in the field and cause uncertainties in the test results. Sources of stray current should be eliminated while measurements are performed, or other techniques need to be used. A brief introduction to the principles that control electrical resistivity and EMI methods is provided in the next sections.

Electrical Resistivity Method

Electrical resistivity methods are commonly applied in the field. The field measurement of resistivity is similar to the laboratory method described above but at a different scale. In a simple arrangement, four electrodes are coupled to the ground via stainless steel stakes and aligned linearly to obtain data (see Figure 6.1). Two of the electrodes induce electric current in the subsurface (which generates an electric field), and two electrodes measure the resulting voltage potential. Electrical resistivity in the field typically is reported in ohm-meters, as opposed to ohm-centimeters reported in laboratory units, due to the relatively larger magnitude of resistivity of some subsurface materials, such as porous rocks (e.g., limestone). The average depth of each apparent resistivity data point is a function of the arrangement of the electrodes at the surface, most notably the distance between the current and

---

[1] Soils in earth retention systems are considered nonaggressive if they have resistivities of more than 3,000 ohm-cm; and soils around non-marine piles are considered nonaggressive if they have resistivities of more than 2,000 ohm-cm (AASHTO, 2002).

FIGURE 6.1 Electrical resistivity testing being conducted in the field. Two of the electrodes placed in the ground induce current (Dipole 1) and the other two measure the voltage potential (Dipole 2).
SOURCE: K. Fishman, committee member.

voltage pairs (see Figure 6.2). The term "apparent" is used to denote that each measurement assumes that the earth is homogeneous. When only four electrodes are used that are relatively close together (i.e., less than 0.5 m apart as in Figure 6.1) and only one apparent electrical resistivity measurement is collected, this assumption is reasonable for the relative volume of soil tested, and the apparent resistivity is assumed to be the true electrical resistivity. Although soil is a heterogeneous material, electrical resistivity is a bulk measurement that cannot discern small changes in the three phases of soils. However, other surveys using many electrodes test a much larger volume of soil, in which case each apparent electrical resistivity measurement likely includes much more significant changes in the three phases (i.e., several different soil layers or the groundwater table). For these measurements, the apparent resistivity cannot be assumed to equal the true resistivity and thus they require more extensive modeling. These multielectrode surveys can also be used for three-dimensional models.

Various complementary laboratory and field resistivity methods have been established (e.g., ASTM G57-20, 2020; ASTM G187-18, 2018) and theoretically should yield the same resistivity. However, given that the moisture content of the sample as received in the laboratory will be different than in situ, it is often difficult to correlate laboratory and field resistivity measurements (see Box 6.4). Some field data acquisition equipment allows a linear array of more than four electrodes and multiple voltage measurements at once, reducing data collection time when the objective is to cover a larger area. The data then are converted to a two-dimensional plot of apparent resistivity, and the true resistivity distribution of the subsurface is obtained through an inversion process. Note that there are inherent limitations to directly using multielectrode field resistivity measurements to determine soil corrosivity,

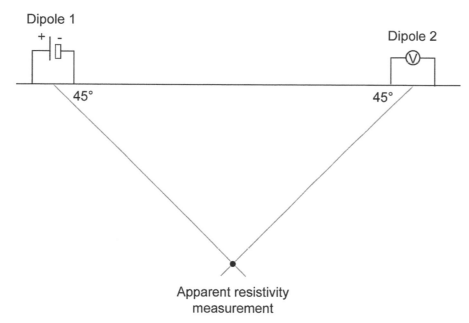

FIGURE 6.2 Electrical resistivity testing array indicating apparent resistivity measurement underground. Dipole 1 is a current electrode pair and Dipole 2 is a potential voltage electrode pair.

BOX 6.4
Relating Laboratory and Field Resistivity Methods

Loehr et al. (2016) observed that field resistivity "should be interpreted considering the in situ water content, recognizing that resistivity will generally decrease with an increasing water content." Adkins and Rutkowski (1998) demonstrated this fact by comparing field- and laboratory-based resistivity measurements on samples of mechanically stabilized earth wall fill at 12 sites in New York. These measurements included minimum laboratory soil resistivity on saturated samples (AASHTO T 288-12, 2016) and in situ measurements of electrical resistivity sounding using the Wenner four-electrode method and the Schlumberger probe arrangements (ASTM G57-20, 2020). Good correlations between field and laboratory test results were obtained only if the moisture content of the laboratory specimens was consistent with the in situ conditions. Karim et al. (2019) supported this finding using field electrical resistivity measurements collected with more electrodes over a larger area and supplementary electrical resistivity tests on soil boxes at varying degrees of saturation.

---

particularly with depth, as the granularity of the inversion process required to interpret field electrical resistivity measurements does not represent the same discrete volume of soil measured in the laboratory. Although application of both finite element and applied element methods to data inversion can reduce artifacts and excessive smoothing, neither technique is perfect. So, while field methods for determining subsurface electrical resistivity exist and are increasingly popular in geo-civil industries, they are not a common standard practice for evaluating corrosivity.

Electromagnetic Induction

Electrical conductivity (or its inverse, electrical resistivity) can also be measured via EMI, a noncontact (i.e., inductive) geophysical method in which the mutual impedance between two or more coils at or above the ground surface, or downhole, is measured. The general method begins with a time-dependent electric current flowing in a transmitter coil, which generates a transient primary magnetic field. This field expands outward, and part of it will flow into the subsurface, which generates an electromotive force and causes eddy currents to flow (see Figure 6.3). The eddy currents generate a secondary magnetic field, which is influenced by the characteristics of the subsurface, namely, the subsurface conductivity. The receiver coil senses both the primary and secondary magnetic fields. The bulk apparent electrical conductivity is determined from the measured primary and secondary fields and is typically reported in millisiemens per meter (mS/m; 1 mS/m = 1,000 ohm-m). Inversion modeling of EMI data is necessary to obtain true electrical conductivity values, but this is rarely done. EMI data are typically used to quickly identify changes in bulk conductivity to indicate, for example, where further investigation is warranted or excavation might prove useful. The apparent depth of the measurement is estimated from the conductivity of the soil, the EMI sensor frequency, and the sensor height above the ground. The term "apparent" indicates that the depth is not a true measurement but rather an estimated measurement based on the assumption that the subsurface is homogeneous.

EMI measurements are commonly used by the oil and gas pipeline industry or for other applications where continuous data along long linear structures are desirable. The commercially available, noncontact devices used for EMI allow for fast and easy data collection over relatively large distances. For example, a person carrying the device can walk with (see Figure 6.4) or drag a sensor via a sled behind a vehicle and set the device to collect data linked with global positioning system locations every few seconds along a pipeline corridor and rapidly collect continuous apparent electrical conductivity measurements.

Assessing Impact of Stray Currents

As discussed in Chapter 2, stray electrical currents may exist around direct current (DC) mass transit facilities, electrical transmission systems, waterfront structures in saltwater, cathodic protection systems, or welding shops,

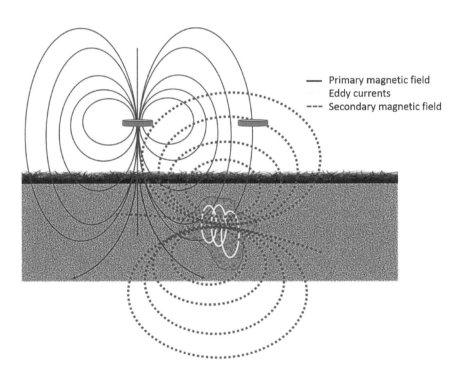

— Primary magnetic field
Eddy currents
--- Secondary magnetic field

FIGURE 6.3 Schematic of an electromagnetic induction surveying system. A transmitting coil generates a primary electromagnetic field (blue) that passes through the ground. A secondary electromagnetic field is induced as the primary field passes through conductive materials. The receiver coil measures the magnitude of the primary and secondary fields.
SOURCE: Saquib Mohammed Haroon, modified with permission.

FIGURE 6.4 An electromagnetic conductivity survey allows workers to walk several miles of pipeline in a day. SOURCE: Mersedeh Akhoondan, committee member.

and may affect buried utility pipes and cables, underground storage containers, and reinforced concrete structures, particularly in high-density urban areas. In general, stray currents rapidly decrease in magnitude away from their source and are not a risk 100 to 200 feet away from the source. Examples of characterizing subsurface stray currents include identifying the distribution of potential gradients between the current source and buried steel of interest. The source might be stray-current leakage from, for example, DC-powered transit systems (Sankey and Hutchinson, 2011). The gradient is a function of rail track-to-earth potential (i.e., between the track and earth) and resistance and can be estimated by measuring earth electrical gradients near the source of the stray current. In contrast, the interference of aerial high-voltage alternating current (AC) power lines on colocated (i.e., nearby) pipelines can be assessed using properties such as the proximity and spatial arrangement of steel and high-voltage AC traveling through the power line and soil resistivity. Finneran et al. (2015) developed a screening tool to identify the conditions under which more resource-intensive analyses and modeling related to stray currents are warranted. Table 6.6 provides simplified examples of the factors, thresholds, and severity rankings they identified; the table also provides some values to consider. Finneran et al. (2015) provide far more detail regarding these and other factors. Their screening tool identifies the severity of risk of high-voltage AC interference with colocated pipeline based on different combinations of factors such as those listed in Table 6.6.

As with any characterization methodology, there is a certain amount of uncertainty associated with the characterization of stray currents and their possible impacts. The difficulty characterizing the subsurface for corrosivity has been described above, and each method, such as determination of resistivity, has its own uncertainties. Cumulative error may be generated using a screening tool such as described by Finneran et al. (2015) because of the uncertainties associated with each factor contributing to the risk assessment. When using such tools, it is important to understand how the tools were developed (e.g., are they empirically or model based?) so that reasonable judgment may be applied with respect to the significance of the results obtained from them. Further complicating characterization of how stray currents affect buried steel infrastructure, if all stray currents and their sources are even known, is the fact that the owners or operators of stray-current sources (e.g., power companies or rail services) will likely be different than the owners or operators of the steel infrastructure. It could be difficult obtaining, for example, the exact location or amperage of the stray-current source.

TABLE 6.6 Examples of Threshold Rankings of Severity of Interference of High-Voltage Alternating Current Power Lines and Pipelines

Example Factors and Thresholds	Ranking of Severity of Interference
Separation distance from high-voltage alternating current source	
Less than 100 feet	High severity
Greater than 1,000 feet	Very low severity
Power line current	
Greater than 1,000 amps	Very high severity
Between 500 and 1,000 amps	High severity
Less than 100 amps	Low severity
Soil resistivity	
Less than 2,500 ohm-cm (laboratory measurement)	Very high severity
Greater than 30,000 ohm-cm (laboratory measurement)	Low severity
Colocation length	
Greater than 5,000 feet	High severity
Less than 1,000 feet	Low severity
Colocation crossing angle	
Less than 30 degrees	High severity
Greater than 60 degrees	Low severity

NOTES: These are examples considered by Finneran et al. (2015). Note that they considered how the combination of factors at specific threshold should trigger more rigorous analysis.

Assessing the Risk of Microbially Influenced Corrosion

Some projects, including large bridge projects, have attempted to use microbial culture (growth) testing to estimate the viable populations of specific bacteria in a given environment. However, these tests are generally unsuitable for assessing the risk of future microbial attacks given the ubiquity of some microbes and the inability to cultivate all species. Instead, projects are more commonly assessed for the likelihood of microbially influenced corrosion (MIC) using subsurface properties. However, those properties that make an environment particularly susceptible to MIC are still debated, and some soil properties are even a consequence of microbial activities. Agarry and Salam (2016) reported that low redox potential and poor drainage due to fine grain size (e.g., clay and silt) were favorable to sulfate-reducing bacteria (SRB), whereas Li et al. (2001) determined that acid-producing bacteria (APB), total organic carbon, resistivity, and water content (listed in descending order of importance) were important. In contrast, Jansen et al. (2017) concluded that the most important properties were sulfate reduction (redox potential, organic matter, and sulfate), mass transfer, and changing conditions (introduction of oxygen or nitrate).

Several standard practices require additional corrosion protection when the subsurface is deemed to be particularly susceptible to future MIC. The Association for Materials Protection and Performance (AMPP) and the American Petroleum Institute (API) require the implementation of specialized measures to prevent MIC when the soil has sulfate concentrations in excess of 500 parts per million (NACE Committee TEG 187X, 2019). AMPP and API also provide some guidance regarding cathodic protection for underground steel piping when MIC is likely (NACE SP0169, 2013[2]). In contrast, ASTM International recommends that redox potential measurements be used to determine the propensity for MIC (ASTM G 200-20, 2020). It states that soils with negative redox potentials are considered severely corrosive, whereas soils with redox potentials exceeding 100 mV are considered noncorrosive. Despite these standard practices, assessing the likelihood of MIC remains difficult because stratified biofilms can often create highly localized conditions that may not reflect highly averaged field surveys of heterogeneous subsurfaces or even samples of soils taken for laboratory analysis. Additionally, the type of infrastructure may affect the susceptibility to MIC (see Box 6.5). Cathodic protection, common on oil and gas pipelines and some water

---

[2] Section 6.2.1.4.1 states, "When active MIC has been identified or is probable (e.g., caused by acid-producing or sulfate-reducing bacteria), the criteria listed in Paragraphs 6.2.1.2 and 6.2.1.3 might not be sufficient."

pipelines, alters the soil environment at the soil–pipe interface that may influence the types, numbers, and activities of soil microorganisms. In addition, many of the legacy coatings used on oil and gas pipelines are biodegradable. Leaks in water distribution systems, which may be ignored from a safety perspective, can contribute oxygenated, treated water to otherwise dry, anaerobic pipe–soil interfaces.

STANDARD PRACTICE DURING OPERATIONS AND AFTER FAILURE

Subsurface characterization after the initial site investigation is not routinely performed as part of operation and maintenance of facilities with buried steel within the geo-civil industries. Observations of external conditions may be documented over time by maintenance personnel or during broader inspection programs during which wet areas, vegetation growing from unintended places (e.g., the face of a retaining wall), rust stains appearing on the facing of a wall system, or poor drainage conditions might be observed. Such observations could inform decisions related to maintenance operations (e.g., to improve drainage) and the need for further evaluation (e.g., exposing buried steel to directly observe conditions) or rehabilitation. Additional subsurface characterization may be undertaken after unexpected metal loss has been observed on exposed steel. This may occur during a retrofit or as construction proceeds to modify an existing facility but is not performed as routine maintenance and operations. Older buried steel may be exposed as an existing facility is demolished or excavated as part of an improvement project. Investigations to determine the cause of observed accelerated corrosion are often conducted, and these include identifying subsurface conditions and measurements of soil properties and characteristics including salt content, resistivity, pH, organics, and moisture content. Samples are obtained from different depths and locations within the subsurface along the facility. Bulk samples from near the surface (to depths of approximately 5 feet) may be obtained from test pits, and samples from deeper depths may be obtained via an auger equipped with a split spoon sampler or in some cases by coring though the face of an existing retaining wall.

More substantial subsurface characterization will occur after a failure. Failure can take various forms but includes observations of unexpected movement of an MSE wall, vertical movement of a bridge pier, soil movement as evidenced by a soil surface depression, or in some cases, catastrophic failure of a structure or portion of a structure. In these cases, there will likely be a forensic investigation into the probable cause of an observed issue, which may include characterization of the current conditions of the underlying soil of a supported structure or soil fill behind a wall. The failure of embedded steel elements due to corrosion should be explored as a possible reason for a failure. When available, geotechnical information from original construction can be useful in examining possible changes in the subsurface environment that may have led to corrosion. As part of a forensic subsurface characterization, soil samples with depth will be taken. This is typically accomplished using standard geotechnical sampling techniques. Unexpected types of soil or fill materials of significant depth or layer could be a sign of inappropriate dumping of materials during original construction. During sampling, the height of the

---

BOX 6.5
Forensic Assessment of Microbially Influenced Corrosion in Pipelines

Microbially influenced corrosion (MIC) of oil and gas pipelines is a recognized international problem, whereas the significance of MIC in the corrosion failures of cast iron and carbon steel drinking water pipes remains controversial. Spark et al. (2020) reviewed the topic of pipeline breaks in drinking water pipes (both cast iron and carbon steel) in Australia. Breaks were due to localized corrosion at the pipe and soil interface, attributed to the heterogeneous nature of soils and microorganisms in the soil. In contrast, Melchers (2020) concluded that MIC was not a problem for buried cast iron pipes in drinking water systems because microbial growth in soils (in general) was limited by nutrients (e.g., nitrogen, phosphorus, and sulfur). The uncertainty of the importance of MIC in water pipelines may partially be derived because a recommended test method (NACE TM0106, 2016) is routinely used for diagnosing MIC in the oil and gas industry, whereas there is no recommended test method for buried steel pipes in water distribution systems.

water table should be noted to understand possible environmental factors that can be associated with corrosion. Soil samples can be tested to characterize corrosivity by measuring resistivity. Chemical analysis of contaminants in the soil such as chlorides, sulfates, or other corrosive species should be performed to measure concentrations of these corrosive species.

Because of the damage that can occur from failures, many oil and gas pipelines are required by regulation to periodically monitor and inspect for the integrity of the asset. While many technologies exist to help identify the areas most likely to need repair or remediation (see Chapter 7), the relatively shallow placement (usually about 3 feet deep) of pipelines allows for direct inspection or forensic investigation. As such, these pipeline "digs" will often include obtaining a soil sample at the point of contact with the pipeline and any associated corrosion identified to characterize any constituents that may have exacerbated the condition. Resistivities, chlorides, sulfates, and microorganisms are examples of the types of properties that are evaluated to determine whether cathodic protection and a reapplication of protective coating is needed.

Standard Practice for Characterizing Microbially Influenced Corrosion During Operations and After Failure

Detailed test methods for diagnosing MIC after it has occurred are available (NACE TM0106, 2016), but early detection of MIC is difficult. MIC can be monitored using coupons that can be easily removed and visually inspected. The presence of slime, odors, and pitting may be indicative of MIC (Little et al., 2020). Coupons can be removed and further examined using other diagnostic methods. MIC diagnostic methods rely on quantifying specific groups of microorganisms, especially bacteria, or some constitutive property (e.g., deoxyribonucleic acid [DNA], ribonucleic acid [RNA], and adenosine triphosphate [ATP]). DNA can be isolated from active and inactive cells and can remain intact after cells are dead. Both RNA and ATP are constituents of active cells.

The traditional method for detecting groups of bacteria associated with MIC is the serial dilution-to-extinction method. This uses culture media designed to grow specific microorganisms (e.g., SRB or APB) (Little and Lee, 2007). Tenfold dilutions of either solutions or buffered suspensions of solids (e.g., soil or corrosion products) in diagnostic media are used to estimate population sizes. Growth is detected as turbidity or a color change, which indicates a chemical reaction. The method is based on the premise that a single cell will produce a reaction in the culture medium. Consequently, small numbers of cells can be detected. However, large numbers of cells require multiple dilutions. The goal is to dilute to the point that bacteria are no longer detectable in a sample (i.e., extinction). Estimated population size is reported as the highest dilution to produce a positive result (e.g., a dilution of 10^3). Commercial test kits for use in the laboratory or field are available with explicit directions for sample collection, inoculation, and interpretation. All culture media are selective to specific organisms and do not account for the nonculturable organisms. Kieft (2000) estimated that only 0.001 to 4 percent of soil microorganisms could be cultured on organic growth media. Therefore, results from culture techniques can provide misleading results—both false positives and false negatives. Furthermore, most commercially available test kits are designed to enumerate bacteria and do not account for fungi or archaea. These respective groups can influence corrosion of carbon steel by organic acid and sulfide production, respectively.

To address issues related to culture techniques, more recent MIC diagnostic efforts have focused on molecular microbiological methods. Hybridization techniques (e.g., fluorescence in situ hybridization [FISH] and DNA arrays) use probes (i.e., DNA fragments designed to bind to DNA or RNA from specific microorganisms [Shakoori, 2017]). FISH probes can be designed to target total bacteria, archaea, or specific groups (e.g., SRB). Only active cells are stained with the fluorescent dye. In contrast, polymerase chain reaction (PCR) can be used to synthesize and amplify a specific section of DNA with a "primer" or "probe" (e.g., targeting the 16S ribosomal RNA gene; Klindworth et al., 2013). Following PCR, the amplified regions are sequenced and used to identify microorganisms. However, unlike hybridization techniques, PCR does not distinguish between DNA from live and dead cells. This is problematic because extracellular DNA (e-DNA), which is DNA that occurs outside of living organisms, may be the largest fraction of total DNA in some environments. Extracellular DNA and DNA from dead cells indicate that there is no direct link between a characterized DNA pool and actual living microbial cell abundance and diversity in a sample (Eid et al., 2018; Makama et al., 2018). Reverse transcription of RNA followed by PCR or quantitative PCR is a PCR-based approach to quantify active cells. However, the design of probes and primers in the PCR techniques requires prior knowledge about the microorganisms in the sample. Most important, numbers of bacterial cells as determined by any method and MIC kinetics have not been correlated. In some circumstances, low numbers of microorganisms could be responsible for severe corrosion, and in other circumstances, large numbers of the same organisms may not be related to corrosion (Little et al., 2020). Therefore, microbial quantification only indicates the extent of microbial presence but does not allow for estimation of the role of microorganisms in corrosion processes.

EMERGING PRACTICES: USE OF DECISION SUPPORT TOOLS

Many laboratory standards evaluate the properties of soils using a specific size fraction. For example, a standard test for resistivity developed by AASHTO (AASHTO T 288-12, 2016) is performed on the fine-grained sample fraction that passes a 2-mm sieve (i.e., No. 10 sieve). The test focuses on fine-grained material because electrical current will travel along paths where lower-resistance fine material is concentrated, meaning that corrosivity is controlled more by the properties of the finer portions than by the bulk portions of the material. Therefore, the test only accurately reflects the subsurface when the sample collected has a significant fraction of fine-grained material. Measurements of corrosion rates (estimated from weight loss and thickness measurements) and resistivity from North American and European sites are available from a database catalogued as part of National Cooperative Highway Research Program 24-28 (Fishman and Witham, 2011). Fishman et al. (2021) culled the Fishman and Witham (2011) data such that coarse samples with less than 22 wt % smaller than 2 mm were removed from the dataset. The culled data are presented in Figure 6.5a compared with data that are not discriminated based on the coarseness of the sample in Figure 6.5b. The report concluded that AASHTO T 288-12 (2016) is only appropriate for measuring resistivity for materials where greater than 22 wt % of particles are less than 2 mm. For material where less than 22 wt % of particles are smaller than 2 mm, other standards that test the material in its "as-received gradation" were considered more appropriate (e.g., Arciniega et al., 2021).

Fishman et al. (2021) attempted to improve methods for characterizing the corrosivity of earth materials by creating a proposed protocol for sampling, testing, and characterizing earth materials (described in flowcharts in Figures 6.6 and 6.7). This simple decision flowchart can be used to determine if the procedures described by Arciniega et al. (2021)—samples tested in "as-received gradation"—or the AASHTO procedures (samples tested after separation on a 2-mm [i.e., No. 10] sieve) will be more accurate given the gradation of the sample. The flowchart begins with determining the particle sizes of the material (i.e., soil gradation). Flowcharts and decision-making tools have been used for many years, including in NCHRP Report 408 (Beavers and Durr, 1998) and AASHTO R 27 (2001), which describe decision support system (DSS) tools for making decisions about data collection, design, and monitoring for corrosion of steel piles, as well as NCHRP Report 477 (Witham et al., 2002). The latter provides a recommended practice for service-life modeling of ground anchorages. More detailed variations of such DSSs could be developed and made available in the future to guide decisions in a variety of corrosion-related practices. One example of a Web-based DSS is offered through the Geo-Institute of the American Society of Civil Engineers.[3] GeotechTools is a decision-making tool that identifies suggested technologies for specific construction applications using project information and constraints, such as the type of application and the purpose of the technology. For example, a selected application of "construction on unstable soils" for the purpose of "compaction" on GeotechTools.org results in a number of suggested techniques (traditional compaction, intelligent compaction, rapid-impact compactor, or high-impact rollers) and presents information about the establishment or maturity of that technology.

ASPIRATIONAL BIOCEMENTATION METHODS
FOR CONTROL OF ENVIRONMENT

Biocementation is an aspirational technology that could impact corrosion of buried steel structures. Bacteria-mediated calcium carbonate ($CaCO_3$) precipitation is a ground improvement technique that requires an external

---

[3] See https://www.geoinstitute.org/geotechtools/login (accessed June 21, 2022).

(a)
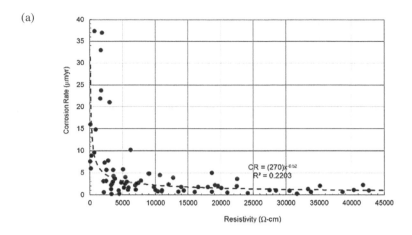

(b)
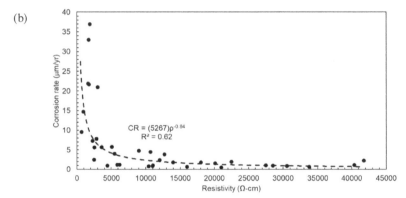

FIGURE 6.5 Measurements of resistivity and galvanized steel corrosion rates (estimated from weight loss and thickness measurements) from worldwide data. **(a)** Testing with AASHTO T 288-12 (2016) from including samples with less than 22 wt % passing the 2-mm sieve. **(b)** Testing with AASHTO T 288-12 (2016) from samples with more than 22 wt % passing the 2-mm sieve. The curve represents whether **(b)** better fits the data, indicating that the characteristics of the materials should indicate which test to use for a given circumstance.
SOURCES: Fishman and Withiam (2011); Fishman et al. (2021).

source of calcium and is designed to decrease the permeability and increase stability of soils (Mujah et al., 2017; Umar et al., 2016). Some bacteria, particularly those that break down urea (ureolytic bacteria), influence the precipitation of $CaCO_3$ by the production of the urease enzyme. Hydrolysis of urea by urease produces carbon dioxide (CO_2) and ammonia (NH_4), increasing the pH and precipitating $CaCO_3$. Biocementation has been used in a variety of geotechnical engineering applications (Ferris et al., 1996; Whiffin et al., 2007) in both sandy (Nemati and Voordouw, 2003) and organic soils (Sidik et al., 2014). Biocementation coats or bridges individual soil particles, gradually reducing the pore size within the soil fabric, and reducing the hydraulic conductivity. Despite the relative efficiency of ureolysis compared to other possible pathways for biocementation, there are some challenges, including environmental challenges associated with the high production of ammonia during the breakdown of urea (Al-Thawadi, 2011). However, as research continues, methods such as this might prove viable for precipitating a corrosion-resistant coating on buried steel at scales relevant to buried steel infrastructure.

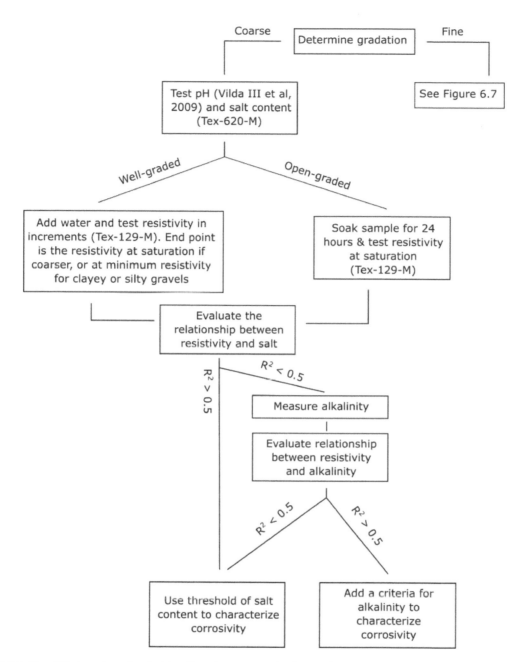

FIGURE 6.6 Simplified version of a decision flowchart uses grain size to dictate test methods and interpretation. This figure describes the decisions related to coarse-grained materials. Figure 6.7 describes the decisions related to fine-grained materials. SOURCE: Adapted from Fishman et al. (2021).

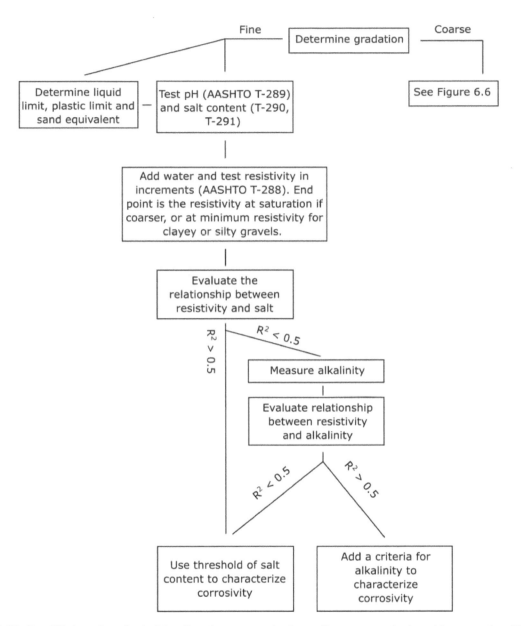

FIGURE 6.7 Simplified version of a decision flowchart uses grain size to dictate test methods and interpretation. This figure describes the decisions related to fine-grained materials. Figure 6.6 describes decisions related to coarse-grained materials. SOURCE: Adapted from Fishman et al. (2021).

7

Standard and Evolving Monitoring Practices

Chapter 6 describes initial subsurface site characterization and post-steel-installation subsurface monitoring. This chapter focuses on practices for monitoring the steel itself. Once in the subsurface, directly monitoring steel corrosion is complicated by lack of direct access to the steel, but some monitoring is possible and is, in fact, required in the oil and gas pipeline industry. The U.S. Code of Federal Regulations (C.F.R.) 49 C.F.R. Parts 192 (gas) and 195 (liquids), as enforced by the Pipeline and Hazardous Materials Safety Administration, requires potential surveys (to be performed annually) to determine effective levels of cathodic protection (CP) going to the entire system (see pipe-to-soil under Potential Surveys to Locate Defects, below). In addition, the code requires that the CP test points (see Chapter 5) must be inspected in-person every 2 months. These checks validate that the current levels determined during the annual survey are being maintained. Other tests described in this chapter, such as in-line inspection (ILI) and aboveground (i.e., over-the-line) potential surveys, are also commonly employed.

There are no regulations related to monitoring buried steel in the geo-civil or water pipeline industries. As does the oil and gas pipeline industry, the geo-civil industries consider safety paramount, but the geo-civil industries address corrosion in the subsurface characterization and design stages. Corrosion monitoring after steel installation is not a standard practice and is done in few instances. Corrosion monitoring for water pipelines is conducted through direct and indirect condition assessments. Many utilities perform direct condition assessment of critical transmission pipelines every few years, but the assessments are not guided by regulation. For pipelines greater than 36 inches in diameter, direct (internal) assessment can include dewatering so that workers can enter pipelines to inspect the lining and corrosion defects and collect removable steel samples. Indirect corrosion assessment techniques—including potential measurements—are also important. Potential surveys can be completed directly by connecting to the pipe through the test stations or pipe appurtenances (e.g., air/vacuum valves) or by indirect measurements using two reference electrodes.

The difference in monitoring approaches among oil and gas pipelines, water pipelines, and geo-civil industries is related to the distinct spatial scales of infrastructure (feet [meters] versus miles [kilometers]), redundancies inherent to the design of the different systems, and the philosophies of corrosion protection versus corrosion minimization, as discussed in Chapter 2. Corrosion prevention, as a primary philosophy of the oil and gas pipeline industry, implies minimization of corrosion and vigilant monitoring. The geo-civil industries, on the other hand, expect a certain amount of corrosion over the infrastructure lifetime. The steel is designed with excess volume to compensate for metal loss expected over the infrastructure design life. Monitoring is not regulated. While corrosion allowance is the prevailing philosophy in the geo-civil industries, it may not be a best practice. The recognition of these philosophical differences is valuable in that it opens discussion about how practices could be improved in the future.

COMMON MONITORING TECHNIQUES

Monitoring techniques range from mostly field-based to mostly laboratory-based, destructive[1] or nondestructive, and fairly simple to extremely complex. One of the simpler techniques is weight or thickness loss measurements in the laboratory of test samples of metal extracted from the field. These measurements can be used to calculate the average rate of corrosion over the time of exposure. Those samples can also be used for other variably destructive laboratory-based methods including microscopic and chemical evaluation. Nondestructive methods include the measurement of the sample corrosion potential (variably referred to as the corrosion potential, open-circuit potential, free potential, or half-cell potential) and techniques such as linear polarization resistance (LPR) and electrochemical impedance spectroscopy (EIS). In contrast, other methods, such as potentiodynamic polarization (PDP), are destructive to the sample because they apply potentials vastly different than the corrosion potential. These approaches are described in this chapter. A laboratory investigation might employ several methods, usually starting with the least destructive (sample corrosion potential, LPR, and EIS) and ending with the most destructive (PDP). Such a combined approach provides different types of data and might take 2–3 hours to complete.

Field-based measurements are beneficial as they provide in situ measurements that can often be performed repeatedly over time. These methods include the measurement of in situ sample corrosion potential, as well as more complex methods that use sophisticated field-based ILI devices (in the case of pipelines). The devices can measure multiple parameters related to external corrosion while traveling in the pipeline.

Monitoring techniques are chosen based on the problem to be addressed. Often, the intent is to determine the location of corrosion or to gather data from which the corrosion rate of the metal in an environment may be estimated. The details of the monitoring method must be carefully controlled as laboratory sample preparation and the sample area can affect the measurement. Table 7.1 lists various methods of monitoring corrosion of buried steel infrastructure, and the subsequent text describes each of those in more detail. The table and text are ordered based on methodologies that primarily use extracted steel samples from the field for laboratory measurements (e.g., direct metal loss measurements, sample corrosion potential, LPR, PDP, and EIS), and then those that are primarily used in the field (potential surveys and ILI).

Direct Metal Loss Measurements (Coupon and Test Sections)

Direct assessment of corrosion of buried steel can be made by measuring the weight, cross-sectional dimensions, or tensile strength of steel prior to burial and then again after some time underground. The mass of steel or protective metal coating loss can be measured and an average corrosion rate can be estimated by comparing those measurements. For example, a test section from a corrugated metal pipe might be obtained by exhuming and cutting a sample from the pipe. The samples are taken back to the laboratory for measurements, and those measurements are compared with preconstruction values. Alternatively, to avoid the need to destructively sample infrastructure, extractable samples of steel of composition similar to that in the infrastructure—called "coupons" (see Figure 7.1)—or other types of inspection elements may be installed directly on or buried near the steel infrastructure during or post-construction and then later exhumed (e.g., Singh, 2014). Visual inspection of the samples may indicate what kind of corrosion has occurred (e.g., pitting). Only average corrosion rates may be calculated, and no information is available about variation over time (e.g., corrosion rates that were initially higher and then attenuated with time; see Chapter 8). Use of coupons is one of the oldest methods for site-specific corrosion monitoring. However, once coupons are exhumed, they are unavailable for future measurements.

---

[1] This report defines a "destructive" test as one that cannot be performed in the field because it damages the infrastructure, or a test that cannot be performed more than once in the laboratory without changing the sample and rendering further tests unreliable. Laboratory tests require the removal of a section or inspection element from the steel on which to perform the test; however, that is not considered "destructive" in this report.

TABLE 7.1 Techniques for Monitoring Corrosion of Buried Steel in Geo-Civil and Pipeline Exteriors

Method	Destructive or Nondestructive	Laboratory or Field Based	Measurement	Application	Industries Used	Efficacy/Issues/Limitations
Direct metal loss measurements (coupon and test sections)	Destructive for coupon	Exposure occurs in the field, but analyses occur in the laboratory.	Visual observation; mass loss (steel or coating); pitting; cross-sectional area loss; tensile strength reduction.	Determine average corrosion rate	Geo-civil and pipeline	Direct simple measurement; extraction of coupons from buried field environment is difficult; get average rate over time of exposure; need multiple coupons to get trend.
Linear polarization resistance (LPR)	Nondestructive	Mostly laboratory based, but field-based LPR is also possible.	Measuring applied current during application of potential to get polarization resistance, which is the slope of potential versus current.	Determine instantaneous corrosion rate	Rare in geo-civil and pipelines	Need Tafel slopes to get corrosion rate, but repeated measurements can be made to get trend; need trained personnel for interpretation.
Potentiodynamic polarization (Tafel slope extrapolation plots)	Destructive	Laboratory based	Measure and extrapolate or fit slopes of cathodic or anodic sections of the polarization curve (potential plotted versus log of current density).	Determine instantaneous corrosion rate and tendency for passivity or pitting corrosion	Rare in geo-civil	Repeated measurements not recommended; strictly only valid for activation polarization; need trained personnel for interpretation.
Electrochemical impedance spectroscopy (EIS)	Nondestructive	Mostly laboratory based, although emerging practices can be applied in the field.	Measuring applied current during application of alternating current potential signal with varying frequency to get polarization resistance from detailed analysis.	Determine instantaneous corrosion rate	Rare in geo-civil	Provides mechanistic information; repeated measurements can be made to get trend; need Tafel slopes to get corrosion rate; need trained personnel for interpretation.
Electrical resistance probes	Destructive for probe	Field based	Increasing resistance owing to loss of cross section of test component.	Determine metal loss and corrosion rate	Emerging in geo-civil	Often results in noisy data that are difficult to interpret.
Potential measurements	Nondestructive	Field and laboratory based	Difference in potential between the surface of a test element (working electrode) and a reference electrode.[a]	Correlated with zinc loss; condition of galvanized reinforcements	Geo-civil and pipelines	Do not provide measure of corrosion rate.
In-line/direct inspection	Nondestructive	Field based	Magnetic flux leakage (MFL) and ultrasonic measurements.	Determine defect size and shape	Oil and gas and water pipeline	The MFL tools are generally very heavy and difficult to deploy. Both methods require close contact with the metal surface and the surfaces should be generally dry (e.g., dewatering of pipe is needed). Both MLF and ultrasonic testing are unable to measure local stresses.

[a] See Box 7.1.

BOX 7.1
Reference Electrodes

When measuring corrosion potential or half-cell potential, a voltmeter is used to measure the potential (voltage) between a structure and a reference electrode in contact with an electrolyte (soil). Reference electrodes have stable, established electrode potentials. Saturated calomel electrodes, silver–silver chloride, copper–copper sulfate reference electrodes (CSEs), and saturated mercury–mercurous sulfate reference electrodes are used in laboratory corrosion studies. CSE are more robust than the others and are commonly used for field measurements of the corrosion potentials of buried structures. When reporting potentials, it is essential to include the type of reference electrode used in the measurement. Attributes of reference electrodes and possible interferences have been described by Ansuini and Dimond (1994).

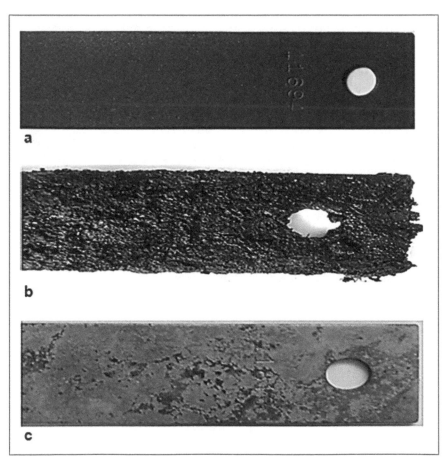

FIGURE 7.1 (a) A new carbon steel coupon prior to burial, (b) a coupon retrieved from a saturated soil, and (c) a coupon after retrieval and cleaning. The difference in the amount of steel between (c) and (a) indicates the amount of corrosion. Average corrosion rates can be calculated by dividing the amount of steel lost to corrosion over the time buried.
SOURCE: Modified from Oparaodu and Okpokwasili (2014). CC BY 4.0.

Given that exhumation is necessary, sampling is typically restricted to near-surface applications. Direct sampling of buried steel is destructive for the test sample, labor intensive, and expensive, and therefore often impractical or impossible. As a result, samples might only be collected if corrosion is suspected. Coupons and other inspection elements are commonly used in geo-civil applications, and water and oil and gas pipelines. In the oil and gas pipeline industry, coupons provide grounding pathways that can interfere with CP systems. Box 7.2 provides an example of the use of inspection elements by the California Department of Transportation. Numerous ASTM standards guide procedures for measuring corrosion on both coupons and test sections. These methods, most commonly applied in the geo-civil industries, are used primarily to estimate corrosion rates for the purpose of modeling design. Methods that rely on electrochemical measurements can be utilized for long-term study to generate an instantaneous corrosion rate.

Potentiodynamic Polarization

PDP is a standard laboratory-based electrochemical method (ASTM G59-97, 2020) that is not used in the field for monitoring corrosion but is included here for completeness. In this method, the potential of the specimen or working electrode is controlled using an instrument called a potentiostat, along with counter and reference electrodes, and the current response is measured. The potential is usually scanned at a slow rate—from a value of about 250 millivolts (mV) below the corrosion potential (E_{corr}; see Box 2.1) to a value of about 250 mV above E_{corr}—and

---

BOX 7.2
Caltrans Inspection Elements

The California Department of Transportation (Caltrans) began installing inspection elements (see Figure 7.2.1) similar to coupons in mechanically stabilized earth walls constructed in 1987 (based on recommendations by Jackura et al. [1987]). Ten-foot-long cold-drawn wire inspection elements are placed within walls using welded wire mat reinforcements and 10-foot-long straps inserted within walls using strap-type reinforcements. The inspection elements are accessible from the wall face and, with the attachment of a hydraulic jack, can be extracted. These inspection elements can also be wired for electrochemical testing. Caltrans physical measurements include observations of surface pitting and loss of section and measurements of remaining zinc for galvanized steel and loss of tensile strength. Inspection elements are spatially distributed in clusters of 18, and long-term plans include extracting three elements each for observations after 5, 10, 20, 30, 40, and 50 years of service. Backfill samples are retrieved from monitoring locations, and data collected during construction are available from stockpile samples and from nearby stations.

FIGURE 7.2.1 Inspection rod being extracted through the wall face with a center-pull jack.
SOURCE: K. Fishman, committee member.

the current is measured. In the United States, the data are commonly plotted with log of the current density (current divided by exposed area) on the *x*-axis and potential on the *y*-axis. An idealized PDP plot is shown in Figure 4.2, which also shows how the curve is analyzed to determine the corrosion current density, or the corrosion rate. The curve points to the left at E_{corr}, and the linear regions in the semi-log plot, which are called the Tafel regions, are extrapolated to the E_{corr} to determine the corrosion current density. The slopes of the lines are the Tafel slopes, which are useful for other techniques as described below. Polarization of the specimen to potentials far from E_{corr} can damage the specimen, so the method is considered to be destructive and is not used for monitoring in the field.

Linear Polarization Resistance and Tafel Slope Measurements

The LPR technique is primarily a laboratory-based method used to estimate the corrosion rate of extracted coupons and in-service steel elements. A small-amplitude potential (approximately ±15 mV from an open circuit) is applied to the specimen. The polarization resistance (Rp), defined as the inverse slope of the current–potential curve (see Figure 7.2), can be correlated to the instantaneous corrosion current density (i_{corr}) with the Stern-Geary equation:

$$Rp = \frac{B}{i_{corr}}; B = \beta_a \beta_c / 2.3(\beta_a + \beta_c)$$

Equation 7.1

where β_a and β_c are Tafel slopes (Stern and Geary, 1957). Then, i_{corr}, in units of $\mu A/cm^2$, can be converted into corrosion rates in milli-inches per year (mpy) using the following equation:

$$Corrosion\ rate\ (mpy) = 0.46\ i_{corr}$$

Equation 7.2

As shown in Equation 7.1, the anodic and cathodic Tafel slopes (β_a and β_c) are required to convert Rp (measured from LPR) to corrosion rate. Tafel slopes are often assumed using values from the literature or can be measured by PDP as described above. Beavers and Durr (1998) directly measured Tafel slopes at their studied sites and found that measured Tafel slopes were almost double the literature values for steel in an earthen environment. This study demonstrates that site-specific Tafel slopes need to be determined in concert with LPR measurements. However, even without knowing the exact corrosion rate, measured decreases in Rp from LPR can also signal a concerning increase in corrosion rate and prompt further investigation.

LPR is a laboratory-based measurement that can be deployed in the field to provide a real-time measurement of corrosion rate. However, it is rarely applied in the geo-civil and pipeline industries because there are no requirements to do so and resources are needed to perform the measurements (Flounders and Lindemuth, 2015). LPR has been used in the field to collect data for performance studies that have been incorporated into research projects (e.g., NASEM, 2011) and is sometimes used in forensic investigations. In the field, a field-based three-electrode configuration is used, whereby a current is impressed between two coupons or in-service steel elements serving as the working and auxiliary electrodes, and the surface potential of the working electrode is measured with respect to a copper–copper sulfate reference electrode (CSE; see Box 7.1) (e.g., IEEE Std 81, 2012).

Electrical Impedance Spectroscopy

EIS involves the application of a small-amplitude alternating current (AC) potential signal of varying frequency, usually around the corrosion potential. This small amplitude of applied potential (~10 mV) makes this a nondestructive technique. The method is useful for assessing corrosion rates under coatings and can be used to characterize corrosion rates and mechanisms for steel samples in soil generally in the geo-civil industries. Like LPR, EIS results can be used to determine the Rp, which again requires knowledge of Tafel slopes for accurate determination of corrosion rate. Analysis of EIS data is significantly more complicated than the other methods and is prone to error and artifact. It has been traditionally used mostly in the laboratory and only now is emerging as

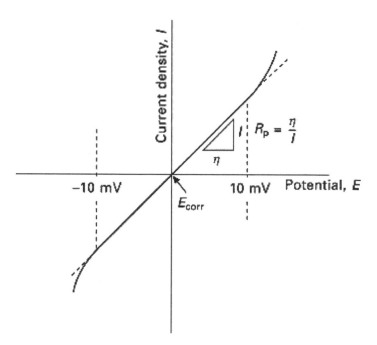

FIGURE 7.2 The inverse slope (polarization potential, η, divided by the current density, I) at the origin of current-voltage plot derived from linear polarization resistance is equal to the polarization resistance (Rp). Rp can be converted to instantaneous corrosion current using the Stern-Geary equation, which can then be related to corrosion rate.
SOURCE: Hiromoto (2010).

a field monitoring approach. EIS provides more information than LPR, and artifacts in the LPR and PDP methods (e.g., ohmic potential drops) are explicitly evident in EIS data.

Potential Surveys to Locate Defects

In many corrosion studies, the measurement of the corrosion potential refers to the potential drop at the electrode interface measured relative to a reference electrode (see Box 7.1). These measurements provide no direct information on corrosion rate, but they can be used to characterize the corrosion process or monitor for change. Although this technique is applied in the laboratory, such measurements are also made in the field in the oil and gas pipeline industries. The pipeline industries often refer to such measurements as structure-to-earth, structure-to-electrolyte, or pipe-to-soil. The corrosion potential testing techniques can be applied to both non-cathodically protected pipe to identify areas that are actively corroding and to cathodically protected pipe to check the performance of a CP system (described in the next section). The CSE (see Box 7.1) is the most used reference electrode for corrosion of steel buried in soil and is cited in most standard practices (SPs), test methods, and recommended practices regarding CP.

One potential survey common in the geo-civil industries is often referred to as half-cell potential and is standardized for reinforcing steel encased in concrete (ASTM C876-15, 2016). This method also has been used occasionally as a nonstandard, nondestructive test method in the field to assess the likelihood of corrosion of buried steel in soil. However, the potential measured in soil cannot be correlated directly to ASTM C876-15 (2016) because that standard relates specifically to likelihood of corrosion of reinforcing steel in concrete. Half-cell potential measurement uses a high-impedance voltmeter to measure the potential difference between steel and a CSE in contact with the soil surface. Figure 7.3 shows the components used in measuring the potential. More negative potential values can indicate where active corrosion may be occurring. The test does not provide corrosion rates.

Other configurations and testing conditions are used commonly in the water pipeline industry, including cell-to-cell and side-drain potential surveys. These can be used to understand and locate corrosion, particularly for non-electrically continuous pipes (AMPP TM0497-2018-SG, 2018; Bianchetti, 2001; NACE SP0169, 2013). Cell-to-cell potential and side-drain surveys measure a potential gradient in the earth using two CSEs and a high-impedance digital voltmeter. The reference cells are repositioned approximately every 5 feet along the pipe. The measurements are stored along with the estimated pipeline station and global positioning system coordinates. In a cell-to-cell survey (see Figures 7.4 and 7.5), one electrode is placed over the pipeline's centerline while the other trails at 5-foot (1.5-m) intervals in a "leapfrog" fashion. Both CSEs are moved along the alignment simultaneously, and the resulting earth potential gradient is measured. The potential gradient measure is related to corrosion or cathodic protection current flowing through the soil to or from the pipe. For this survey, typically the forward CSE is connected to the positive terminal of the voltmeter, and the rear CSE is connected to the negative terminal of the voltmeter. Both references are moving above the center of the pipe simultaneously every time measurements are taken. On the cell-to-cell potential plots, more electronegative potential indicates anodic locations.

In side-drain surveys (see Figures 7.6 and 7.7), one reference is positioned above the center of the pipe (connected to negative terminal of voltmeter) and the other one (connected to positive terminal of voltmeter) is positioned parallel to the pipe at approximately 5- to 10-foot offset (the offset distance is kept constant). When the survey is conducted, both electrodes are moved along the alignment simultaneously, and the resulting earth potential gradient is measured. The side-drain measurements are examined for magnitude and polarity changes to determine the anodic locations.

Alternative potential survey methods in the pipeline industries include alternating current voltage gradient (ACVG) surveys, historically known as Pearson surveys. These surveys are applied to non-cathodically protected infrastructure. The surveys apply an AC signal to the pipeline and measure signal leaks to the soil under areas of coating failure. This leakage of signal generates a potential gradient that can be measured using two "ski poles" or A-shaped frames. A more common variation of this survey is the direct current voltage gradient (DCVG) surveys, which can also identify coating defects but using the current impressed by the CP unit (see Figure 7.8).

A similar method, often referred to as pipe-to-soil potential, is commonly used in the oil and gas industry to monitor CP effectiveness. During effective application of CP to a pipeline, the pipe potential is reduced from the corrosion potential to decrease the corrosion rate. Therefore, the pipe-to-soil potential measured when CP is being applied is not the corrosion potential. Pipe-to-soil potential measurements are completed on test points that

(b)

(a)

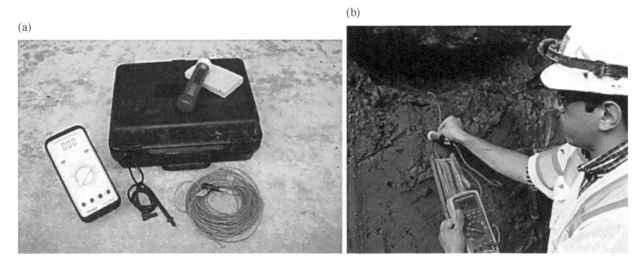

FIGURE 7.3 (a) Potential testing equipment, including copper–copper sulfate electrode and sponge, lead wires, and voltmeter going clockwise. (b) Half-cell corrosion potential testing measurements along the vertical soil surface of an embedded steel pile. The wire running into the hole is connected to the steel infrastructure. The measured difference in potential along the soil surface over a short length of the steel pile was about −450 mV. After removal of the soil, visual observations showed that the most severe corrosion, in terms of metal loss, was at the location of the greatest negative potential.
SOURCE: Pivot Engineers.

FIGURE 7.4 Cell-to-cell survey where the copper–copper sulfate electrode is placed over the pipeline's centerline. SOURCE: Mersedeh Akhoondan, committee member.

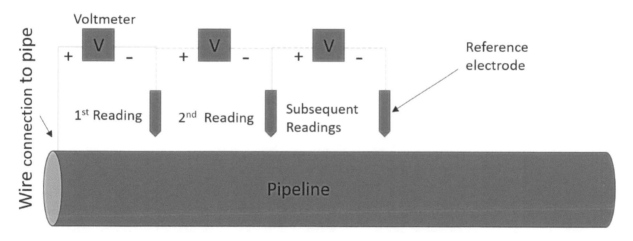

FIGURE 7.5 Schematic of a cell-to-cell survey where the copper–copper sulfate electrode is placed over the pipeline's centerline.

FIGURE 7.6 A side-drain survey where one copper–copper sulfate electrode is positioned above the center of the pipe and the other one is positioned parallel to the pipe at approximately 5–10 feet offset.
SOURCE: Mersedeh Akhoondan, committee member.

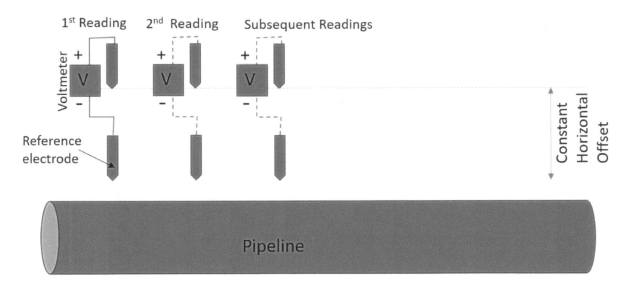

FIGURE 7.7 Schematic of a side-drain survey where one copper–copper sulfate electrode is positioned above the center of the pipe and the other one is positioned parallel to the pipe at approximately 5–10 feet offset.

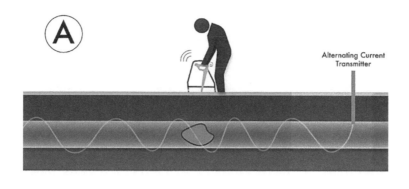

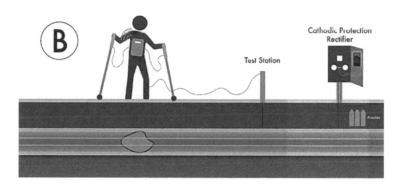

FIGURE 7.8 Schematic of (**A**) alternating current voltage gradients and (**B**) direct current voltage measurements.

are usually about a mile apart along a pipeline and utilize a one- or two-electrode indirect methodology based on electrochemical principles to characterize the condition of the steel, the protective coating, or the effectiveness of the CP system. Surveys are conducted along the length of the steel infrastructure and measure changes in the way current flows over the surface of buried steel. Those test points are connected to a rectifier and ground bed containing anodes for the CP system. The rectifier supplies the current needed to protect the pipeline and the test points measure that an adequate amount of current is reaching those prescribed points (see Figure 7.9). The spacing of rectifiers depends largely on the resistivity of the soil and the condition of the coating. Newer pipelines can have rectifiers spaced over dozens of miles, whereas older pipelines can have them much closer.

NACE SP0169 (2013) provides three criteria for CP of steel and gray or ductile cast iron: empirical evidence that a particular cathodic polarization is adequate, a structure-to-electrolyte potential (voltage) of −850 mV versus CSE, or a minimum cathodic polarization of −100 mV versus CSE. This SP is used as a standard in many state regulations describing required maintenance and monitoring procedures for pipeline operators. Potential surveys on CP infrastructure are not simply a measurement of pipe polarization (η) but also include voltage drops due to soil and pipe resistance (IR; see Figure 7.10). These voltage drops are artifacts in that they do not reflect the potential at the sample–soil interface, so they must be accounted for to assess whether measured voltages across a structure-to-electrolyte (soil) boundary are adequate. "Instant-off potential" is a standard technique for assessing CP because a potential measurement collected immediately after cathodic polarization has been interrupted (instant-off) eliminates the influence of IR. The extent of cathodic polarization is determined by the difference between the instant-off potential and the native potential measured in the absence of cathodic polarization.

Adequate CP is achieved at differing levels of cathodic polarization and depends on the environmental conditions. A single polarization voltage or voltage difference may not provide corrosion control for all environments. When microbially influenced corrosion is identified or probable, a polarized potential of −950 mV versus CSE (or more negative) or as much as 300 mV of cathodic polarization might be required (NACE SP0169, 2013). Similar recommendations are suggested for steels maintained at temperatures >40°C and for mill-scaled steels.

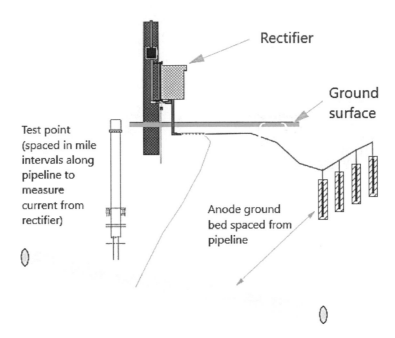

FIGURE 7.9 Visual depictions of rectifiers, anode ground beds, and test points for a cathodic protection system on an oil and gas pipeline. Rectifiers are often tapped from a high-voltage power distribution system and connected to the anode ground bed and the pipeline. Spacing is dependent on current requirements of the pipeline. Test points are located roughly every mile along the pipeline and are used to check that an adequate amount of current is being applied.

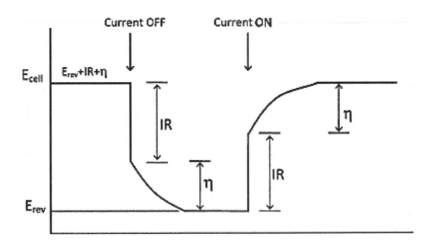

FIGURE 7.10 The potential of an ideal electrolytic cell upon current interruption. When the current is interrupted, the voltage drop (IR) vanishes instantaneously, whereas the polarization (η) decreases slowly to the natural or reversible potential (E_{rev}). When the current is turned on, the voltage drop is immediately apparent, followed by the slow polarization.
SOURCE: Mohandas and Fray (2011).

There are several pipe-to-soil potential methods that utilize current interrupters to measure CP effectiveness. By far the most common are close interval potential surveys (CIPSs), which measure the effectiveness of CP and possible coating defects along stops at regular distances along the pipeline (see Figure 7.11). Although CIPS is not a routine prescriptive, code-required monitoring test, the test provides a detailed survey of the CP system at points every 2.5 feet, as opposed to every mile at test points, and has therefore become an effective and accepted tool for meeting the generic code requirements of "adequately monitoring the CP system" while also providing useful data in conjunction with ILI tools to locate active corrosion on the pipeline.

In-Line Inspection

As described earlier, direct inspection of pipes can be accomplished by people traveling through a drained pipe (common in the water pipeline industry) to take measurements. ILI also can be accomplished using ILI machinery, called "smart pigs," as is common in the oil and gas pipeline industry. The focus of ILI is often identification of interior corrosion caused by fluids transported within a pipe, but measurements of wall thickness also reflect exterior corrosion (the focus of this study). A number of metal loss measurement techniques can be applied during ILI. For example, magnetic flux utilizes a temporary applied magnetic field to identify anomalies in the magnetic flux distribution. An uncorroded wall returns a homogeneous magnetic flux distribution, whereas a corroded wall shows a heterogeneous distribution. The movement of the magnetic flux from corroded areas is referred to as "leakage," and the technique is called magnetic flux leakage (MFL). Transverse MFL is a similar method except the orientation of the magnetic field is turned 90 degrees. While traditional MFL will generally see defects and corrosion of the circumferential variety, transverse MFL can be used to assist with tight axial slotting defects, such as those found in a pipe weld seam. Axial slotting defects can include selective seam corrosion. Ultrasonic methods are another commonly applied ILI technique that measure pipe wall thickness to determine metal loss. In this technique, ultrasonic signals are emitted toward the pipe, and the corresponding echoes are received from both the internal and external pipe surfaces. The time of echo return and the speed of the ultrasound in the applicable steel must be measured or known. As with all ultrasonic technologies, this method requires a couplant—a material applied to the metal to facilitate the sound wave transmission. Therefore, it is a tool more commonly used in liquid pipelines. Some gas pipelines run this tool with a "slug" of liquid couplant to facilitate inspection. A third commonly applied technique in ILI is electromagnetic acoustic transducer, or EMAT, where the transmission of a magnetic field created by transmitter coils is measured by one or more receiver coils. This technique allows for external and internal defect measurement. This technology is mainly used for tighter, planar- or crack-like defects rather than volumetric defects such as a corrosion defect.

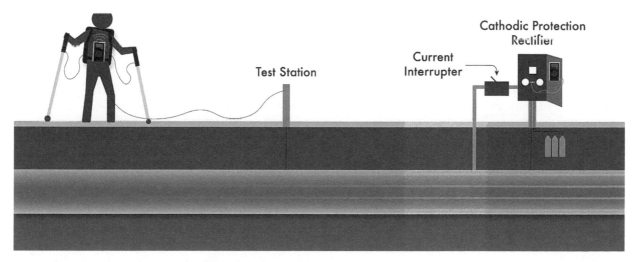

FIGURE 7.11 Schematic of close interval potential surveys.

Because of its relative ease of use and data provided, ILI technology is used for a majority of the oil and gas transportation pipelines in the world to verify their integrity. New pipelines are purpose-built to run this technology, and older pipelines are modified to accept it—a cost-intensive process. Unfortunately, there are still limitations within the technologies, usually associated with their basic physics. Therefore, selection of the right tool for the job can be as important as obtaining the data these tools provide. For example, transverse MFL tools are the standard tools used to detect corrosion anomalies on a pipeline. However, there can be limitations to how the resulting data are evaluated, particularly with respect to much larger areas of corrosion. Transverse MFL uses two sensors at a relatively short distance from each other to determine if there is an "upset" in the conditions on the pipeline wall. If the sensors are moved into a large area of corrosion, both sensors can lose the reference point of the nominal wall thickness and the data can produce a smaller anomaly and make the corrosion appear less severe than it actually is. Conversely, a ultrasonic tool, which provides a straight wall thickness measurement, does not have this issue. As described above, transverse MFL also has difficulty recognizing tighter axial anomalies for which a circumferential MFL tool can better assist. Other tools described above have come into the market more recently to address cracking in pipelines. This technology is evolving as well with time and use.

In the water pipeline industry, a number of different ILI or direct inspection methods are used, including remote-field electromagnetic scans, broadband electromagnetic probes, remote-field transformer coupled scans, MFL scans, in-pipe acoustic velocity wall thickness, and ultrasonic techniques. These also can be performed by a human or remotely using a smart pig. Other tools can be installed on the exterior of pipe after trenching (AWWA M77, 2019).

EMERGING (ASPIRATIONAL) OPPORTUNITIES

Although all industries have standards for nondestructive, field-based methods, they are most commonly used by the oil and gas industry. This may be because of the relative ease of access in oil and gas compared to the geo-civil industries (e.g., accessing a deep foundation is challenging compared to a relatively near-surface pipe and the access granted by the pipe interior). Because of the relatively recent push in the geo-civil industries for increased asset management, it is anticipated that nondestructive, field-based monitoring will become more routine.

Electrochemical Impedance Spectroscopy in the Field

Laboratory electrochemical measurements are performed on isolated test steel components because it must be known over which area the current flows. It is difficult to know exactly where the current is flowing in the field or larger uncontrolled settings. However, frequency-domain electrochemical methods are possible as field methods and are an area of active research in applied geophysics (e.g., Hördt et al., 2007; Kemna, 2000; C. Wang et al., 2021). There is often confusion because the nomenclature is different in different fields (i.e., corrosion engineering, geophysics, biomedical engineering), and the nomenclature changes based on the experimental setup. For example, in geophysics the analogous corrosion laboratory EIS measurement is called spectral induced polarization (SIP) when the frequency is from 0.1 to 1,000 Hz. Field-based SIP measurements are often called complex resistivity or complex conductivity. Furthermore, higher-frequency, above 10^4 Hz, measurements may be called dielectric spectroscopy or complex permittivity. Note that these are common frequency ranges. Ultra-broadband electrical spectroscopy has been measured on soils from 10^{-3} to 10^9 Hz in the laboratory (Loewer et al., 2017).

Regardless of the frequency range, making measurements in the field remains aspirational because current instrumentation requires electrodes to initiate a signal and measure the resulting voltage potential of the subsurface. Insulated wires are needed to connect these electrodes to the instrumentation, and extensive insulated wiring is needed to obtain clean signals. Working with such extensive wiring systems can be challenging in the field (Binley and Slater, 2020). Furthermore, where low-frequency measurements are of interest, collecting field data becomes difficult given power needs or simply the amount of time the technician requires to perform the measurement. Truffert et al. (2019) recently addressed this limitation using time-synchronized transmitter and receiver waveforms.

Field EIS measurements are more commonly obtained in the time domain, at one frequency and coupled with electrical resistivity—known as induced polarization. Tucker et al. (2015) used induced polarization to evaluate the depth of unknown bridge foundations, including steel piles, by measuring the response of the subsurface around the pile. In this case, a corroded steel pile will be relatively more chargeable than a noncorroded pile in the same soil. It may be possible to use this method to identify corrosion of buried steel; however, as Hubbard et al. (2003) note, the method is largely qualitative. It can identify where there is a change in the measurement but not necessarily the cause. In addition, SIP has also been used to monitor calcite precipitation rates during biocementation (see Chapter 6). However, this application is related to the control of the environment and not directly related to monitoring the corrosion of steel (Kessouri et al., 2019).

Finally, as previously mentioned, it is not possible to know exactly where the current is flowing when using the techniques in the field. For this reason, near-surface geophysicists rely on inversion methods to interpret data. Inversion methods result in nonunique solutions and are designed to provide a general indication of an area of interest, not necessarily an exact solution that one obtains in the laboratory.[2] Still, novel inversion techniques such as the analytical element method (Steward, 2020) to overcome previous mathematical limitations are being explored to improve inversion methods and provide more robust and discrete field data.

Electrical Resistance Probes

The electrical resistance of a solid body depends on its dimensions. Therefore, it is possible to estimate the corrosion rate of a metallic element by measuring the metal's change in resistance resulting from the loss of material by corrosion. Electrical resistance (ER) probes measure the change in electrical resistance of steel relative to reference noncorroding steel within the probe body. At present, the technique is only applied in the field after a major failure or when extreme corrosion is anticipated (see Figure 7.12).

The resistance R of a body with length L and cross-sectional area A is given by the following:

$$R = \rho \, L/A$$

where ρ is resistivity. As corrosion occurs from the outside in, the metallic cross section of the body decreases and R increases, so the change in R provides a means to estimate weight loss. However, resistivity is a sensitive function of temperature; therefore, temperature changes need to be accounted for when assessing resistance changes. This is accomplished by incorporating a "dummy coupon" as a control into the ER probe body that is protected from the soil but not from temperature changes. Changes in resistance in the coupon due to changing temperature are measured, and those resistance changes can be accounted for in the resistance measured on the subject steel infrastructure. ER probes can provide a sustained measurement of metal loss, which can be used to calculate corrosion rate without having to remove the probe from the environment. Furthermore, the approach is not electrochemical in nature so it can be used in environments where electrochemical measurements are difficult or impossible, such as in poorly conducting electrolytes. Figure 7.12a shows a schematic of a monitoring system installed at select piers at the Leo Frigo Memorial Bridge (Wisconsin) after there had been a pile foundation failure at one pier. As can be seen, the system includes both half-cell potential and ER probes.

In Situ Coupon Measurements

Some industries install coupons that are used for in situ measurements taken at regular intervals during service. This is useful because corrosion rates often decrease with time (Romanoff, 1957), and time-series measurements are necessary to assess the rate of steel loss and strength decay. The New York State Department of Transportation, the Kentucky Transportation Research Cabinet Engineering Services, the Florida Department of Transportation, and the Georgia Department of Transportation perform nondestructive tests including measurement of half-cell potential and LPR on in-service reinforcements and coupons installed and wired for monitoring. The North Carolina Department of Transportation routinely installs coupons during construction of mechanically stabilized earth walls and monitors in-service reinforcements and coupons using half-cell potential to evaluate the remaining zinc and service-life for their galvanized reinforcements.

---

[2] See, for example, recorded presentation to the committee by Mark Everett, available at https://www.nationalacademies.org/event/03-09-2021/laboratory-and-field-geotechnical-characterization-for-improved-steel-corrosion-modeling (accessed June 24, 2022).

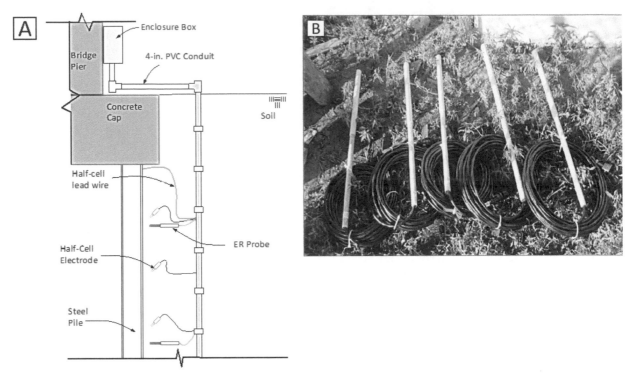

FIGURE 7.12 **(A)** Schematic of monitoring system installed at Leo Frigo Memorial Bridge (by Pivot Engineers). **(B)** Before installation of these electrical resistance (ER) probes, the white polyvinyl chloride (PVC) pipe near the top of the picture is removed to expose a metallic probe.
SOURCE: Pivot Engineers.

Fiber-Optic Strain Sensing

Significant recent work in infrastructure monitoring has focused on the use of fiber optics as a technique that can deliver high spatial resolution of several physical properties. Fiber-optic sensors can be embedded within infrastructure, on its surface, or around the infrastructure to measure temperature (distributed temperature sensing), strain (distributed fiber-optic strain sensing), acoustics (distributed fiber-optic acoustic sensing), and even pressure (distributed fiber-optic pressure sensing). Corrosion of buried steel results in a loss of cross section of the material, which can lead to measurable strain. These measurements are collected at a desired temporal interval that can range from frequent sampling over brief periods to stable measurements for years to monitor infrastructure. Using the distributed fiber-optic technique, engineers can obtain real-time measurements to monitor in situ conditions or damage related to changes in these properties. Distributed fiber-optic techniques are commonly used for structural health monitoring in aboveground infrastructure, such as bridges, and there have been many related laboratory studies. Zhao et al. (2011) used fiber-optic sensors to monitor the corrosion of steel in reinforced concrete in the laboratory. Distributed fiber-optic sensors have also been used for subsurface infrastructure monitoring (Pelecanos and Soga, 2017) but not directly related to corrosion. In pipelines, these sensors are used for perforation monitoring, leak detection, pipeline integrity monitoring, and strain deformation modeling (Sasaki et al., 2019). As these methods have proven successful in the laboratory for monitoring corrosion and in the field for monitoring strain of buried steel and buried reinforced concrete infrastructure, they are an aspirational method for monitoring corrosion of buried infrastructure.

8

Predictive Modeling

Corrosion is a process that, in theory, could be accurately predicted if sufficient information about the corrosion/infrastructure system is available. An engineering model, in general terms, is a simplified physical or mathematical representation of a system that can be used to predict system behavior and engineering performance, and can be used to influence engineering design. Models differ in complexity and scale, but they may be used to better understand how characteristics of the steel or subsurface will affect metal loss or inform choice of corrosion protection. The corrosion process is dependent on the details of the corroding interface including surface and environment composition and structure. Variations in these factors at the micrometer scale can have a controlling influence on corrosion and can also change with time. However, it is rarely possible to obtain sufficient information to accurately predict exactly when, where, or by what mechanism corrosion will occur. Ultimately, the goal in the field of the corrosion of buried steel structures is related to understanding the reliability of the structures. Community experts in structural reliability have developed methods to estimate the integrity and reliability of structures (e.g., Melchers and Beck, 2018). However, the effects of corrosion are generally not considered in detail (see Box 8.1).

The modeling approaches used by the geo-civil and oil and gas pipeline industries differ greatly, as do other aspects of engineering practice. The geo-civil industries tend to use models prior to design to predict metal loss with time and therefore to determine the additional steel necessary in design to compensate for that metal loss due. The oil and gas pipeline industries tend to emphasize modeling of the steel infrastructure during operations and maintenance, using data collected from indirect or in-line inspection (ILI) monitoring (see Chapter 7). Table 8.1 provides examples of current and evolving model types. The models are classified as deterministic and nondeterministic and are listed in order of increasing complexity and computational resources. Deterministic models are those that do not include randomness and will always produce the same output from given starting conditions. Nondeterministic models, in contrast, can exhibit different behaviors on different runs from the same starting conditions because these models consider different possibilities among the variations of input parameters (e.g., soil properties, initial zinc thickness, metal loss model parameters). Models used for geo-civil applications are generally deterministic and often involve empirical or semiempirical estimates of metal loss. These empirical and semiempirical approaches historically have dominated in corrosion engineering due to limited databases and computing power, but nondeterministic models may be better suited to situations that have a large number of variables and uncertainties. Only those conducting academic research and those in the oil and gas pipeline industry commonly use nondeterministic models.

BOX 8.1
Efficacy of Buried Steel Infrastructure Design Methods

The lack of field data from long-term, statistically designed, and well-controlled experiments poses a significant challenge to assessment of the efficacy and uncertainty in current design methodologies. Owing to the complexity of the interactions between the subsurface and steel, a range of different types of predictive models have been developed, from empirical to numerical and deterministic to nondeterministic. However, without high-quality field performance data, validation of these models and quantification of their uncertainty is not possible.

As a result, practice has evolved to use an "upper-bound" method, which is similar to that used in other fields (e.g., the historical approach to liquefaction). In this approach, the design is based on a range of initial subsurface conditions and does not predict changes that may occur over the infrastructure service life. An aggravation factor[a] is applied to account for heterogeneity, changes in subsurface conditions, and other unknown factors, resulting in a conservative upper-bound approach to design. Because most buried steel infrastructure either provides critical services (e.g., fuel transmission) or poses a safety hazard in the event of failure (e.g., bridge foundations, rockslides, or retaining wall failures), conservative design is often warranted and relatively few failures have been observed. For example, out of tens of thousands of mechanically stabilized earth (MSE) walls that are currently in service, approximately a dozen have been documented as failures due to corrosion of the MSE reinforcements (Gladstone et al., 2006, plus more recent failures not yet documented in the literature). All of these cases were correlated with corrosive environments. Unfortunately, conservative design results in large economic cost and still does not preclude failure in all cases. A more fundamental understanding of the mechanisms and their interactions that cause corrosion would allow for less costly construction with less uncertainty in design.

_____

[a] A correction for variables not accounted for in the data from which the empirical relationships are derived (e.g., localized metal loss).

While multiple model types are used in corrosion management practice, the most common, especially in the geo-civil industries, are empirical models. These models assume the same corrosion behavior at sites with similar characteristics and properties. Empirical models are most powerful when informed by laboratory and controlled field test data and combined with observations from in-service performance. However, in practice the data that form the basis for empirical models are limited in terms of breadth, quantity, and quality. Other types of models, including analytical, numerical, and statistical models, also may be used in practice, especially in the oil and gas industries. The next few sections describe models applied commonly in practice related to corrosion management. The final section of this chapter describes emerging trends in corrosion-related modeling.

EMPIRICAL MODELS FOR CORROSION RATE BASED ON SUBSURFACE PROPERTIES

One type of empirical model estimates corrosion rates based on subsurface characterization measurements (see Chapter 6). King (1977) described an American nomogram based on soil resistivity and pH that provides average corrosion rates for burial times of greater than 2 years. That nomogram can be used to predict rates of weight loss over time and corrosion due to pitting. Using corrosion data from field surveys, King checked the accuracy of the nomogram and found that the model was able to predict corrosion rates with ±10 percent accuracy for 30 percent of cases tested, but the remaining cases were accurate to only ±50 percent. He noted that the nomogram generally predicted lower corrosion rates than indicated by field measurements of corrosion, but corrosion rates in sandy soils predicted by the nomogram appeared slightly more reliable. This indicates that this nomogram provides only rough estimates of corrosion rates and is not dependable for use in design. Models similar to this nomogram have been produced by various state transportation agencies for galvanized steel and by NACE (2001). NACE modeled corrosion rates of steel piles above the water table correlating data from Romanoff (1957) and from seven other

TABLE 8.1 Types of Models Applied to Problems Related to Corrosion of Buried Steel

Model Type	Definition	Typical Use	Example Outputs for Corrosion Applications	Example Inputs for Corrosion Applications
Deterministic Models				
Empirical	Directly relate input to a database of observations or experimental data	Research and the geo-civil industries	Corrosion rate	Measured soil resistivity and pH
			Metal loss	Experimentally derived estimates describing corrosion rate in 1 year and a time constant describing change of corrosion rate over time
Semiempirical	Use simplified calculations to relate input to a database of observations or experimental results to obtain an answer	Academia and industry use	Metal loss due to galvanic corrosion	Difference in the respective corrosion potentials; kinetics of the electrochemical reactions on each metal, areas of each metal, and the ohmic resistance of the electrolyte to which they are exposed
Analytical	Mathematical models that describe an exact change in a system through solution of a mathematical analytic function	Used in research and rarely in oil and gas industries	Effectiveness of cathodic protection and the likelihood of corrosion	Measurements of a corrosion defect and of potential, current, and impedance profiles
			Corrosion rate	Measured electrochemical potential at surface, measured or calculated thermodynamic and kinetic parameters
Numerical	Computation of a large number of mathematical equations to find an approximate solution	Used in research and rarely in oil and gas industries	Finite element modeling to predict corrosion anomalies for cathodic protection systems	
			Predictions of galvanic corrosion for components with complex shapes	Electrical or thermal field given conductivity and appropriate boundary conditions
Nondeterministic Models				
Statistical	Mathematical model incorporating set of statistical assumptions about sampling of corrosion processes or corrosion damage	Used in research and in oil and gas industries[a]	Probability of pipeline failure, uncertainty of corrosion rate, uncertainty of corrosion depth	Defect measurements
Stochastic	Mathematical model to estimate spatial/temporal variation of possible outcomes; variation usually based on historical data	Used in research and in oil and gas industries[b]	Spatial distribution or occurrence of corrosion defects or events, temporal fluctuation of corrosion rate or current	Historical inspection data; soil survey results; climate, geological, and geotechnical data

continued

TABLE 8.1 Continued

Model Type	Definition	Typical Use	Example Outputs for Corrosion Applications	Example Inputs for Corrosion Applications
Machine learning: Supervised (artificial neural network, linear regression, support vector machine); Unsupervised (clustering)	Data-driven modeling paradigm for constructing machine learning algorithms	Used in research and emerging in oil and gas industries[c]	Predicting corrosion depth and location with quantified uncertainty, estimating similarity of corrosion environments among different sites	Historical inspection data, soil survey results, domain knowledge, and engineering experiences

[a] Examples include Aziz, 1956; Caleyo et al., 2007; Evans et al., 1933; Gong and Zhou, 2017a,b, 2018a,b; Greene and Fontana, 1959; Gumbel, 1954, 2004; Ji et al., 2017; Mears and Evans, 1935; Melchers, 2003, 2004, 2005a, 2008; Shibata, 1991, 1996; Shibata and Takeyama, 1976; Zhou et al., 2017.

[b] Examples include Bazán and Beck, 2013; Caleyo et al., 2009; Dann et al., 2015; Engelhardt et al., 1999; Hong, 1999; Kamrunnahar and Urquidi-Macdonald, 2010, 2011; Laycock et al., 1990; Maes et al., 2009; Martín et al., 2010; Melchers, 2010, 2015; Provan and Rodriguez, 1989; Rodriguez and Provan, 1989; Shibata, 2013; Shibata and Takeyama, 1977; Valor et al., 2007; Velázquez et al., 2009, 2010; Zhang and Zhou, 2015; Zhang et al., 2013; Zhou, 2010.

[c] Examples include Dann and Birkland, 2019; Kamrunnahar and Urquidi-Macdonald, 2006; Rosen and Silverman, 1992; H. Wang et al., 2015a,b,c, 2016, 2019, 2021; X. Wang et al., 2021; Wen et al., 2009; Yajima et al., 2014, 2015.

sites. NACE (2001) concluded that corrosion rate is a function of pH and resistivity (ρ), with lower resistivity and pH correlating with higher corrosion rates:

$$\text{Corrosion rate} \propto (\text{pH} * \log{(\rho)})^{-1} \qquad \textbf{Equation 8.1}$$

The same relationship does not hold true for galvanized reinforcements because both acidic and alkaline conditions result in elevated zinc corrosion rates (Pourbaix, 1974). The regression presented by NACE is based on limited data, and the correlation is not high. However, it is an example of a multivariate model that estimates corrosion rate based on subsurface properties.

EMPIRICAL MODELS FOR METAL LOSS (ROMANOFF MODELS)

Another type of empirical model is that which uses experimental datasets (e.g., Romanoff, 1957) to derive equations and calculate metal loss over a specific number of years. Romanoff (1957) concluded that buried steel corrosion rates attenuate with time, depending on the degree to which the soil is aerated, which, in turn, is dependent largely on drainage. He found this observation applied to general corrosion as well as localized corrosion from pitting. He observed an approximately linear relationship over timescales of 10 to 20 years after plotting the logarithm of maximum pit depths or weight loss (uniform loss of thickness) versus the logarithm of time, such that they conformed to

$$P = k_1 t^{n1} \qquad \textbf{Equation 8.2}$$

and

$$X = k_2 t^{n2} \qquad \textbf{Equation 8.3}$$

where P and X (both in micrometers) are the maximum pit depth and uniform loss of thickness, respectively, after time t (years); k_1 and k_2 (both in micrometers per year) are the maximum pit depth and loss of thickness after the first year, respectively; and $n1$ and $n2$ are exponents that are less than 1. The values of $n1$ and $n2$ are related to the aeration of the soil, with well-aerated soils corresponding to lower values of $n1$ and $n2$. Well-aerated soils with an abundant supply of oxygen have a high scaling tendency—the oxidation and precipitation of iron as ferric hydroxide occurs close to the metal surface to produce a protective scale. The protective scale formed in this manner tends to decrease corrosion rates with time, and this is modeled using lower values of $n1$ and $n2$. In poorly aerated soils, the products of corrosion remain as ferrous ions with a lower oxidation state and tend to diffuse outward into the soil, offering little or no protection to the buried steel, such that the initial corrosion rates decrease slowly with respect to time, if at all. The formation of a protective scale and attenuation of corrosion rates may also be affected by the corrosivity of the soil. Even in a well-aerated soil, high concentrations of soluble salts may prevent precipitation of protective layers of corrosion products. The rate of corrosion would not decrease over time, and the corresponding value of the time constants would be closer to 1.

An iteration of the Romanoff model was presented by Darbin et al. (1988) and is applicable for galvanized steel in constructed earth fills. This study provides additional data from a 20-year study that evaluated the corrosion of galvanized steel elements in free-draining granular soils. This is important as fewer than 10 percent of the original Romanoff (1957) data come from free-draining granular soils such as those used in constructed earth applications, and not many include galvanized steel samples. Using these data, Darbin et al. (1988) proposed that metal loss of galvanized steel in free-draining fills could be described with a constant exponent "$n2$" equal to 0.65 and a coefficient of k_2 equal to 25 μm/yr or 20 μm/yr when fills have laboratory-measured resistivities of 1,000–3,000 or >3,000 ohm-cm, respectively. The constants k_2 and $n2$ correspond to an upper bound of corrosion data instead of a best fit. In this sense, there is an unquantified margin of safety inherent to the Darbin model (see Box 8.2). The Darbin model is used commonly in the geo-civil industries and forms the basis for metal loss modeling, calculation of the amount of steel needed to compensate for corrosion losses, and simplified forms that consider corrosion rates to be constants over specified time intervals. The accuracy of the model varies with respect to soil parameters, and considerable scatter exists between predictions and observation for higher corrosion rates. In general, the model is applied as an envelope to the data and is conservative. The model cannot be applied to the behavior of soils or metal types that were not included in the original database.

However, there are other mechanisms of corrosion than the uniform and pitting corrosion described by Equations 8.2 and 8.3 (see Chapter 4). The effects of other localized corrosion mechanisms are often incorporated into Equation 8.3 using a "correction factor," also referred to as an "aggravation factor." For example, Darbin et al. (1988) and Elias (1990) incorporated a factor of 2 to the steel corrosion rate after zinc is consumed (Equation 8.5) to account for macrocell corrosion processes in galvanized steel that may occur along with general corrosion. This is expressed as

$$\text{If } t \leq \left(\frac{z_i}{k2}\right)^{1.54}, \text{ then } X = 0 \qquad \textbf{Equation 8.4}$$

but

$$\text{If } t > \frac{z_i}{k2}^{1.54}, \text{ then } X = 2 \times k_2 \times t^{n2} - 2 \times z_i \qquad \textbf{Equation 8.5}$$

where z_i is the initial thickness of zinc coating (μm), X (μm) is the loss of base steel thickness after a certain number of years (t), k_2 is the loss of thickness after the first year (μm/yr), and $n2$ is an exponent that is less than 1. This factor is strictly applicable to mechanically stabilized earth (MSE) wall reinforcements and is not necessarily applicable to describe the corrosion processes inherent to other systems.

A subsequent modification of this model was proposed by Bastick and Jailloux (1992) to describe how the corrosion of galvanized steel is affected by the salt content of fill. They demonstrated that the effects of chloride and sulfate species can be added together to render values of "k_2" as a function of salt content such that

$$k_2 = 0.21(\text{Cl}^-)^{0.86} + 2.74\left(\text{SO}_4{}^{2-}\right)^{0.32} \qquad \textbf{Equation 8.6}$$

where [Cl$^-$] and [SO$_4$] are the concentration of chloride and sulfate ions (ppm) in the fill.

BOX 8.2
Selecting a Margin of Safety

As with any other engineering design analysis, a margin of safety that addresses uncertainty and reduces or controls the likelihood of failure and loss of service needs to be applied within empirical models based on the National Bureau of Standards (NBS) data (Romanoff, 1957). One modeling approach is to characterize the corrosivity of the subsurface conditions for a site and extract portions of the database (e.g., NBS data) with characteristics of the subsurface that are consistent with the observed site conditions. These data are modeled to estimate metal loss and the need to compensate for corrosion with added section of steel. A margin of safety is meant to account for the following:

- **Limited on-site data.** This includes limited on-site soil corrosivity sampling frequency, the variability of the on-site soils, and the risk that sampling will not represent the most soil corrosivity on-site.
- **Limited reference database.** There is a risk that the characteristics of soil at a specific site may not be represented in existing databases.
- **Other corrosion factors.** Local corrosion effects that cannot be easily incorporated directly into models including bimetallic coupling (galvanic corrosion, such as a steel pile and copper grounding), pH concentration cells (e.g., a steel pile in contact with a concrete pile cap), and microbially influenced corrosion (which can occur in certain soil conditions) may affect modeling calculations.

Approaches to incorporating a margin of safety into the design are available including (1) multiplying the required sacrificial steel thickness by a factor (e.g., such as an aggravation factor to account for the possibility of localized corrosion due to the development of macrocells); (2) use of extreme-value statistics to generate models that are an upper bound to the expected rates of metal loss; or (3) applying reliability analysis and selecting metal loss with a given probability of not being exceeded within a specified service life (e.g., 5 percent probability of failure for a redundant system). Selecting the appropriate margin of safety requires engineering judgment facilitated by applying multiple approaches and comparing predictions from different models that are applicable. The engineer selects margins of safety based on the confidence in the representiveness, variability, and frequency of soil corrosivity data used for identifying the relevant entries from a given database and

- confidence in the ability to identify within a database those data that are consistent with site conditions;
- required design life and criticality of the structural element;
- harmonization factors of safety related to corrosion with structural design safety factors; and
- past project experience near the site.

In contrast to the above equations, other models approximate steel loss using a linear extrapolation of a constant steel corrosion rate that is multiplied by design life. One example is the American Association of State Highway and Transportation Officials (AASHTO) model that is specific to the design of galvanized reinforcements in MSE walls with noncorrosive or "mildly" corrosive fill.[1] This model can be calculated as

$$X = (t_f - C) \times k_{2c}; \; C = t_1 + \frac{z_i - (k_{2a} x \, t_1)}{k_{2b}} \qquad \textbf{Equation 8.7}$$

where X is the loss of metal thickness (μm), t_f is the design life (years), C describes the time for zinc depletion, t_1 is the duration for which the initial corrosion rate for zinc prevails (2 years for the AASHTO model), z_i is the initial thickness of zinc, k_{2a} is the initial corrosion rate of zinc for the time (t_1), k_{2b} is the subsequent corrosion

---

[1] According to AASHTO, MSE fill must comply with the following electrochemical criteria: pH = 5 to 10, resistivity ≥3,000 ohm-cm, chlorides ≤100 ppm, sulfates ≤200 ppm, organic content ≤1 percent.

rate of zinc until exhaustion, and k_{2c} is the corrosion rate of steel after zinc depletion. The AASHTO metal loss model defines values for "k_2" at which first zinc, then steel, will be lost from the galvanized steel cross section:

k_{2a}: Loss of zinc during first 2 years ($t_1 = 2$)	15 μm/yr
k_{2b}: Loss of zinc after first 2 years to depletion	4 μm/yr
k_{2c}: Loss of steel (after zinc depletion)	12 μm/yr

Both the AASHTO and Darbin et al. (1988) models compute the same metal loss with respect to a service life of 65 years. But for longer service lives (e.g., 75 or 100 years), the linearized AASHTO model renders more metal loss compared to the power law implemented for the Darbin model (Fishman and Withiam, 2011). The AASHTO model is the most popular metal loss model applied to the design of MSE reinforcements. However, this model should only be applied for fills that meet the limits and ranges of electrochemical parameters specified by AASHTO and is only applicable to galvanized steel.

Subsequent developments that have built off of Romanoff (1957) include those by Melchers and Petersen (2018) in which they reinterpreted the data presented by Romanoff (1957) and proposed a bimodal model for estimating corrosion rates and service life. That bimodal model considers the progression of corrosion and is useful for recognizing the effect of soil type on the time-dependant performance of buried steel.

ANALYTICAL AND NUMERICAL MODELING FOR GENERAL AND LOCALIZED FORMS OF CORROSION

The general corrosion rate of metals in the subsurface can be predicted using mixed potential theory if the kinetics of the anodic and cathodic reactions on the metal surface are known. Mixed potential theory is based on the notion that the rates of the anodic and cathodic reactions must be equal under open-circuit conditions (i.e., no electrical connection to another metal or measuring instrumentation) to uphold charge conservation. The anodic and cathodic electrochemical reaction kinetics are assumed to follow the well-accepted Tafel equation that relates the rate of the reaction to the electrochemical potential at the surface (see Chapter 4) or mass transport limitations. The corrosion potential and rate (current density) can then be determined from the point where the lines representing the kinetics for the two reactions intersect (see Figure 4.2). Thermodynamic (equilibrium potential) and kinetic (exchange current density and Tafel slope) parameters are required as parts of the Tafel equation. As described in Chapter 7, these parameters can be determined empirically, but commercially available software uses thermochemical principles to estimate them. One example of commercially available software is OLI Studio from OLI Systems, Inc. (Anderko et al., 2001). A comprehensive thermodynamic database is used to provide a detailed speciation of the local environment including ion activities, transport properties, and fractional surface coverage of adsorbed species, which then allows for determination of the required parameters in the Tafel equation using other models. The software also contains models for the prediction of passive oxide film formation and the stability of localized forms of corrosion such as pits and crevices.

Localized forms of corrosion can be modeled given advances in the understanding of the physicochemical process involved (Frankel et al., 2017; Li et al., 2018a,b, 2019a,b, 2021; Oldfield and Sutton, 1978). Models for pitting corrosion focus primarily on pit growth. The ability to predict initiation events is less developed. Galvanic corrosion (see Chapter 4) can also be modeled using this approach. Several factors will control the extent of the galvanic interaction: the difference in the respective corrosion potentials, the kinetics of the electrochemical reactions on each metal, the areas of each metal, and the resistance of the soil to which they are exposed. If these factors are known, then it is possible to accurately predict the extent of galvanic corrosion. Furthermore, it is possible to make predictions for components with complex shapes using finite-element approaches to solve the Laplace equation, which describes an electrical or thermal field given a conductivity and appropriate boundary conditions (Palani et al., 2014).

Commercially available software exists specifically for the prediction of galvanic corrosion. Some examples include CorrosionMaster (Elsyca, 2022), Galvanic Corrosion Simulator (BEASY, 2022), Corrosion Djinn (Corrdesa, 2022), and OLI Studio (OLI Systems, 2022). These software tools are used primarily in the research community but are increasingly being used by infrastructure designers to predict corrosion rates and galvanic corrosion susceptibility. The U.S. Department of Defense has just approved a new standard practice (MIL-STD-889D, 2016) that "defines and classifies galvanic compatibility of electrically conductive materials and establishes requirements for protecting electrically conductive materials in a dissimilar couple against galvanic corrosion." The standard replaces an older standard based only on corrosion potentials, which can result in incorrect recommendations. The new standard is based on polarization curves using predictions of the sort provided by the commercial software listed above. These programs require a reliable and robust database and struggle to predict variations in conditions over time, including the metal surface conditions and local electrolyte composition, which can have large effects on corrosion rate.

NUMERICAL MODELS FOR PREDICTING STRAY CURRENT

As described in Chapter 4, stray current is commonly produced by the electromagnetic inductance of a high-voltage alternating current power line, direct current (DC)-powered transit systems, or cathodically protected structures. Numerical modeling can be used to understand variables that affect the magnitude of the stray current (Finneran et al., 2015). For stray current produced from a power line, a transmission line model is used to calculate the propagation direction parallel to the electric field vector. The routing of the steel infrastructure and transmission line networks are incorporated into the numerical model colocations represented with connected finite sections and nodes. The model renders the potential, current, and current density for each colocation. For DC-powered transit systems, numerical models can be used to estimate how well the design of the transit system complies with allowable stray current (Flounders and Memon, 2020). Corrosion engineers can use those data to calculate the stray-current leakages and determine the level of track-to-earth resistance needed in the design. The utility of the software is limited by the accuracy of the input data. Often the uncertainty in critical input variables such as the current load, electrochemical properties of soils and fill, and the effects from transients (e.g., formation of scale from corrosion products) limits the benefits of a more complex model.

MODELING CATHODIC PROTECTION POTENTIAL AND CURRENT PROFILES, POLARIZATION, AND CURRENT DENSITY

Chapter 5 describes how buried steel forms a natural electrochemical system that can be altered to protect the buried steel by installing a cathodic protection (CP) system, and Chapter 7 describes how the CP effectiveness can be monitored. Modeling of the electrochemical processes in these systems is often useful both in design and during monitoring of the CP system. Before CP installation, modeling can help establish the current density necessary to properly protect the steel (typically in units of milliamperes per square foot). This current density may change due to the distance from the CP source and the local subsurface conditions (e.g., resistivity). Another parameter commonly modeled during design is the polarization of the buried steel infrastructure. Subsurface conditions can have a strong impact on polarization, which can change the overall effectiveness of the CP.

After installation of infrastructure, current and potential distribution models are commonly used to monitor cathodically protected steel. The distance from the CP source and the local conditions of the buried structure and subsurface often leads to different current distributions that are measured during monitoring with methods such as close interval potential surveys (CIPSs). Modeling the potential and current profile distribution locates the sites where the steel might not be protected (see Figure 8.1). Kennelley et al. (1993) and others have quantified potential and current distribution of CP-protected pipelines for several conditions, including inhomogeneous soil, with "holidays" or coating defects and with varying anode distribution. These data have been used reactively to adjust CP conditions and respond to corrosion concerns. Commercial software exists specifically for the prediction of CP characterization. Some examples are Comsol (Fontes and Nistad, 2019), Elsyca (2022), and BEASY (2022).

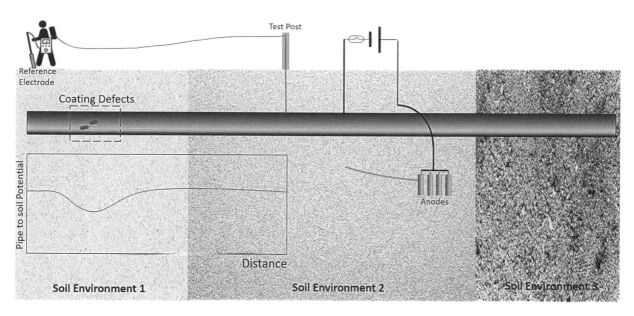

FIGURE 8.1 Potential profile being collected over three different soil environments from a reference electrode connected to a test post indicates regions on the steel with coating defects.

EMERGING AND EVOLVING TRENDS

Many of the emerging or evolving trends in corrosion modeling use probabilistic modeling and machine learning (ML) to calculate statistics instead of relying on factors of safety. However, these types of models often require larger datasets than are currently available. Other emerging trends attempt to use basic materials parameters on the atomic scale to produce models predicting the lifetime of macroscale steel infrastructure. The next sections describe aspects of these modeling techniques.

Probabilistic Modeling for Corrosion Rate and Severity

Recent efforts in corrosion science have attempted to try to implement probabilistic reliability-based models, instead of "go/no-go" models that are based on factors of safety. This has been more successful in recent years in the oil and gas pipeline industries that utilize ILI data from "smart pigs" to measure size, shape, and locations of defects (see Chapter 7). Models can help determine the corrosion rate and whether the defect should be repaired or replaced. One of the earliest and most practical probabilistic models in corrosion science is the burst prediction model (ASME B31G-2012 (R20170), 2017), which takes the measurements of a corrosion defect and determines if and at what pressure the pipe is likely to fail. However, this model does not truly model corrosion but rather models the pressure that can be sustained by a corroded pipe. Additionally, this model uses a factor of safety instead of true probabilistic modeling to determine failure probability. More complicated probabilistic analysis can categorize the severity of the corrosion and the rate of corrosion using multiple different ILI technologies and soil measurements (Yajima et al., 2015). Whereas the technologies utilized in ILI and subsurface characterization are standardized, the analysis and computer tools are not.

Another type of probabilistic model that can be applied to corrosion science is the application of extreme-value statistics, which use statistical approaches to focus on the tails of a distribution. Aziz (1956) was the first to recognize that the deepest pit in a distribution of corrosion pits will be the likely cause of an eventual failure, which led to the application of extreme-value statistics to model such data (Melchers, 2005b,c, 2008; Shibata, 1991, 1994). Asadi and Melchers (2017) applied extreme-value statistics to the corrosion of buried cast iron water pipes. The deepest pits were found to follow a Gumbel distribution, which is a mathematical function used to describe the maximum values in a dataset. Another approach for predicting pitting corrosion is based on comparing the distributions of the corrosion (E_{corr}), pitting (E_{pit}), and repassivation (E_{rp}) potentials. One study (Cong and Scully, 2010) concluded that "pitting can occur once the maximum [open-circuit potential] rises up to within three standard deviations of E_{pit} (99.7 percent) and exceeds E_{rp}."

Machine Learning

Few large and comprehensive datasets exist in corrosion science and corrosion engineering. The Romanoff (1957) data are limited in scope, which reduces the predictive ability of any analysis of his data. However, if a sufficiently large dataset were available (e.g., on the order of thousands of data points), then sophisticated new analysis methods could be applied. In recent years, approaches that simulate human thinking—known generically as artificial intelligence—have been applied in numerous fields such as handwriting and speech recognition, marketing, economics, robot locomotion, and search engines. ML is a subset of artificial intelligence that uses computer algorithms to seek and analyze trends and patterns in datasets and create predictive relationships or assign categories. ML approaches are classified as either supervised or unsupervised. Supervised learning uses a training sample set with predetermined outputs to build a model capable of predicting the output of unknown samples. In contrast, unsupervised learning uses a training set without preassigned outputs or labels with the goal of allowing the algorithm to find patterns and clustering. Effective ML requires large, high-fidelity datasets to allow accurate training as well as validation and testing of the algorithms.

The oil and gas pipeline industries have generated large volumes of data as a result of regulatory requirements for active integrity programs and the ability to collect pipe corrosion data using sensors in smart pigs. Those data combined with the need to maintain the integrity of hundreds of thousands of miles of pipeline have stimulated the use of ML in the pipeline industry. Pipeline operators must assess the risks and consequences of failure (e.g., as a result of a leak). Recent papers have summarized the use of ML in pipeline integrity management (Ossai, 2020; Rachman et al., 2021; Soomro et al., 2022) as well as for the detection of defects in inspection data (Rachman et al., 2021).

Different ML methods exist, some of which are available as simple laptop-based software or libraries (Paszke et al., 2019; Pedregosa et al., 2011). A common approach, artificial neural network (ANN) analysis, attempts to mimic the behavior of the human brain, which is skillful at sensing patterns (Bishop and Nasrabadi, 2006). ANN utilizes a connected array of nodes, similar to neurons, including input nodes (data descriptors), output nodes (desired predictions), and intermediate hidden nodes that connect them. Mathematical expressions form the connections between the nodes. ANNs have been applied to corrosion data in recent years. For example, Birbilis et al. (2011) used ANN to model the effects of composition on corrosion rate and strength of 68 different magnesium alloys. The model predicts the properties of untested alloys, possibly accelerating the development of new alloys with improved properties. Neural networks have also been used to assess images of steel structures to determine, by ML only, if they were corroded (Atha and Jahanshahi, 2018). Almost 70,000 images of both corroded and noncorroded regions were used for training, validation, and testing. The precision was found to be greater than 90 percent.

Other ML methods employ decision-tree analysis and regression approaches such as ordinary linear regression (Weisberg, 2005), ridge regression (Arashi et al., 2019), kernel regression, and logistic regression (which is commonly used in biostatistics to sort out the efficacy of drugs or other medical treatments; Johnell and Klarin, 2007). Bayesian networks are useful for capturing the knowledge of human experts and the rules they use for relating causes and outcomes (Heckerman, 2008). This approach of knowledge-based analytics is particularly useful in areas—such as corrosion—where data are often scarce but experts have deep understanding. ML can also integrate stochastic models with explicit consideration of uncertainty, which is suitable for complex heterogeneous soil distribution environment (H. Wang et al., 2015a,b, 2016, 2019, 2021; X. Wang et al., 2021; Yajima et al., 2015). It has become a tool used in the pipeline industry to characterize and quantify damage, risk, and integrity (Kim et al., 2021).

Whereas the oil and gas pipeline industries have generated large volumes of data and have applied ML techniques, those databases are often proprietary, or their use is restricted for homeland security reasons. ML could be applied more commonly to predict corrosion of buried steel in geo-civil infrastructure given a sufficient dataset representing corrosion rates of buried steel over a range of conditions and including a variety of soil types, groundwater levels, moisture contents, resistivities, pH, and chloride and sulfate concentrations.

Atomistic Modeling and Density Functional Theory

Fundamental first-principles modeling of corrosion has been a focus of research in recent years. One approach, known as density functional theory, solves an equation similar to the Schrodinger equation for a small cluster of atoms to determine factors such as the electron density distribution, system potential energy, band structure, and the equilibrium atomic structure. These parameters can be used for a variety of fundamental applications relevant to corrosion, such as assessing chemical bond strength, reaction mechanisms, activation energies and reaction kinetics, and mechanical properties (Ke and Taylor, 2019). These are basic materials parameters, which would need to be fed into larger-scale models to arrive at practical predictive models of corrosion. The goal is a multiscale model that starts with electronic configuration and ends up with a prediction of a component lifetime. Practical application of atomistic modeling will not be achieved without considerable resources and time.

9

Conclusions and Recommendations

This report describes a number of practices associated with identifying and mitigating corrosion of steel in contact with earth materials throughout the life cycle of infrastructure. The practices do not necessarily represent state of the art. Corrosion protection practices among industries differ, with some industries adopting a corrosion allowance approach (the geo-civil industries) and others adopting a corrosion avoidance approach (the oil and gas pipeline industries). The different practices are often driven by necessity, but inertia can prevent improvements in practices. It is always challenging to adopt new practices based on the best available science and technological developments, and advances in one industry are not translated easily to another. The spatially and temporally complex and heterogeneous subsurface environment makes it impossible to completely characterize the subsurface, and so it is difficult to accurately predict when, at what rates, and by which mechanisms corrosion will occur. Researchers and developers in industry and academe attempt to improve knowledge and practice related to corrosion, but they often are reliant on limited decades-old corrosion-related data (Romanoff, 1957) and random corrosion data referenced to subsurface environments. Additionally, practices are often driven by standards used for efficiency and reproducibility but that might be derived from standards developed for other purposes. Industries and different disciplines within an industry use inconsistent vocabularies and do not routinely share data related to system design and performance.

The oil and gas and the geo-civil industries have different corrosion-related challenges. The oil and gas industries cannot realistically perform comprehensive site assessments for the entire lengths of infrastructure. Corrosion is recognized as an important failure mechanism, infrastructure is designed to protect against corrosion, and corrosion protection systems (e.g., cathodic protection [CP] systems) are monitored. Geo-civil industry infrastructure designs, on the other hand, are driven by the need to withstand loads and hazards such as earthquakes and floods. Far less consideration is given to failure due to corrosion of buried steel. Design in the geo-civil industries is based largely on the expectation that corrosion will occur and either (1) extra material is incorporated into steel cross sections or (2) treatment or protection is added to the steel to inhibit corrosion for the design life of a structure. Failure due to corrosion is not expected in the typical infrastructure design life, and so corrosion generally is not monitored. A vulnerability of both industries is that changing conditions that affect corrosivity, corrosion mechanisms, and corrosion rates are not accounted for in design and are not monitored during service.

This chapter presents a set of recommendations intended to improve understanding of buried steel corrosion, corrosivity of the subsurface, and decision making related to subsurface characterization, infrastructure design, monitoring, and operation and maintenance. The recommendations are necessarily visionary in nature and will require those with interest in and responsibility for corrosion of buried steel to think differently than is common in current practice. The committee contends, however, that visionary approaches are necessary to enable improvements in knowledge and practice related to site characterization, design, and monitoring. Without such improvements, practitioners will continue to rely on possibly overly conservative practices and high factors of safety based on sparse empirical evidence to compensate for high levels of uncertainty and few data to validate models. Such practices regularly result in safe infrastructure but perhaps at higher economic cost than necessary to achieve optimal performance.

There is a common need among industries and specific disciplines to reduce uncertainties at each phase of the steel infrastructure life cycle so that risks can be better understood, modeled, mitigated, and monitored, resulting in better and more efficient infrastructure design, construction, and management and increased welfare and safety for the nation. To address this need, conclusions and recommendations are presented in the following themes:

- consistent terminology and common lexicon,
- multidisciplinary research,
- comprehensive longitudinal experimentation,
- data analytical techniques,
- decision support systems (DSSs),
- opportunistic data collection, and
- development of a data clearinghouse.

Although a number of the recommendations build off each other, they are not provided in any order of importance, and many of the recommendations could be implemented in parallel.

IMPROVED COMMUNICATION THROUGH CONSISTENT TERMINOLOGY

The committee that produced this report did not expect the complex deliberations it experienced; after all, practice in multiple industries is driven by numerous standards and guidelines based (presumably) on knowledge gained from research and experience. However, the committee discovered that its members from different disciplines and industries sometimes spoke using the same words but with different meanings that have been generated within their respective silos of practice. They found that "common knowledge" might not be common to all industries, and it was often supported by too little data. Imprecise language could exacerbate misconceptions about corrosion and corrosivity.

To deliberate effectively, the committee found it essential to define a common vocabulary and to constantly check for consistency of its use throughout the preparation of this report. As examples, "corrosion potential"—an electrochemical term (see Box 2.1)—was used by some committee members to suggest a likelihood of corrosion; the scales at which "pitting" occurs (i.e., the size of what is considered a "pit") was defined differently among committee members; and the definition of "soil" was a source of confusion early in the study process, as was the fact that soil is a multiphase electrolyte of interest consisting of solids, liquids, and gases. Until agreement was reached about terminology to be used in this report, discussions were confusing, and draft text was ambiguous and even contradictory. With common terminology, discussions were more productive, and conclusions drawn from them were made with greater (and more justifiable) confidence.

Just as the committee had to take a multidisciplinary approach in its deliberation, the broader technical communities that address corrosion need to apply a multidisciplinary approach to understand corrosion of steel in subsurface environments. This means that fields such as geotechnical engineering, structural engineering, earth science, materials science, hydrology, metallurgy, corrosion engineering and modeling, geophysics, geochemistry, and microbiology all need to inform a complete understanding of corrosion and subsurface corrosivity. However, there is limited communication among practitioners and researchers between these disciplines and between industries with interest in understanding, detecting, or protecting against the corrosion of buried steel infrastructure. Knowledge is not transferred, and, poor decisions may be made as a result of miscommunication. Benefits could be realized by the larger community of corrosion scientists and engineers, researchers, and practitioners in all the relevant industries, and industry stakeholders—including the public—with improved communication.

Recommendation 1: Standards-making bodies from different industries, in collaboration with the public agencies with responsibilities related to buried steel infrastructure, and researchers interested in understanding and preventing buried steel corrosion should develop a common lexicon with precise definitions associated with corrosion of steel and the characterization and monitoring of subsurface environments in which steel is buried or placed.

More technically precise terminology will better convey issues related to corrosion of buried steel. Increasing communication and collaboration between disciplines and industries would facilitate attainment of common goals such as better understanding of the corrosivity of a subsurface environment, better prediction of corrosion and corrosion rates, and more effective design, construction, and management of buried steel infrastructure. However, the technical communities that need to collaborate often use different vocabularies to refer to the same concepts, or the importance of some concepts is not recognized. A common lexicon is a first step to sharing knowledge and advancing practice.

Agreeing on and using a common lexicon might seem a trivial recommendation, but achieving this goal will be difficult because of entrenched vocabulary usage. Nonetheless, until all use terminology with a common understanding, intellectual silos and silos of practice will persist and opportunities to advance innovations across disciplines or from research into practice will be limited. Professional societies (e.g., the Association for Materials Protection and Performance, the American Association of State Highway and Transportation Officials [AASHTO], and ASTM International) might collaborate to develop this lexicon. Because standards-making bodies already have influence in the public and private sectors, their collaborative development and dissemination of a common lexicon would help ensure that the individual technical communities they serve will incorporate the vocabulary into their work.

MULTIDISCIPLINARY RESEARCH

Engineers who design and conduct site characterization investigations are rarely knowledgeable about corrosion mechanisms, and corrosion engineers are often unfamiliar with complexities of the soil–groundwater–gas electrolyte. Even fewer engineers are familiar with subsurface microorganisms and how their presence influences corrosivity. No individual expert from any sector should be complacent about assumptions regarding the likelihood and occurrence of corrosion of buried steel. Corrosion protection will be more effective with improved understanding of corrosion and the factors that contribute to a corrosive environment rather than routine application of higher factors of safety. Improved understanding will yield better tools and standards, better practices, and, ultimately, more sustainable infrastructure. Given the complexities of the subsurface environment and the numerous factors that contribute to corrosivity, improved understanding of corrosion mechanisms and rates for buried steel will require research undertaken using a different approach than that which has been applied to research to date.

Increased interaction between, for example, corrosion scientists and practicing geotechnical engineers may help practitioners advance from the use of overly simplistic models to more sophisticated modeling techniques to improve corrosion prediction and management capabilities. Box 4.1 describes the successful application of multidisciplinary and multisectoral communication and collaboration during the forensic investigation of the failure of the Leo Frigo Memorial Bridge in Wisconsin. That investigation and identification of the necessary remediation required the combined specialty knowledge of structural and geotechnical engineers, hydrogeologists, geophysicists, CP specialists, corrosion engineers, microbial testing experts, and those with expertise in chemical analysis of fill and soil materials. No single individual could have identified the problem or solutions without the expertise of the others. Box 9.1 provides another example of a large-scale successful multidisciplinary effort.

There is a need for public agencies, industry groups, and academe with interest in or responsibilities related to corrosion of buried steel to formalize collaborative efforts to identify and facilitate multidisciplinary research for improved prediction of, protection against, and monitoring of the corrosion of buried steel. Primary goals of such collaborative multidisciplinary efforts include increasing understanding of how multiple ground conditions

---

BOX 9.1
The Multidisciplinary Center for Earthquake Engineering Research:
A Successful Multidisciplinary Research Initiative

The Multidisciplinary Center for Earthquake Engineering Research (MCEER, previously NCEER), established by the National Science Foundation in 1986 at the University of Buffalo, is a national effort that helps communities develop "knowledge, tools, and technologies to increase disaster resiliency in the face of earthquakes and other events" (University at Buffalo, 2022). Recently, MCEER has expanded its focus from earthquake engineering to any natural or engineered event that is hazardous to critical infrastructure, facilities, and society.

MCEER advances its mission by conducting multidisciplinary research and complimentary education and outreach initiatives. The multidisciplinary research completed by MCEER involves seismologists; geologists; civil, structural, geotechnical, fire, and wind engineers; statisticians; owners/operators; city planners; and oil and gas, nuclear, transportation, aerospace, and electric power industries. Research and development conducted in MCEER are well documented and include studies on infrastructure relevant to this report, including pipelines (e.g., Eguchi et al., 1995; O'Rourke and Liu, 2012), highway structures (e.g., Power et al., 2006), and bridges (e.g., Fishman and Richards, 1997). Additional results from these cooperative efforts have included revised descriptions of earthquake hazards in terms that are more useful to practicing engineers, updated codes and standards, improved methods of analysis, development of performance-based design, a systems-based approach to design that considers multiple hazards and system vulnerabilities, development of seismic protective systems, improved material durability for infrastructure, verification and validation process for numerical models, techno-economic and risk assessment for nuclear facilities, improved communication and data collection activities, and advancement of the state of the art, particularly with respect to earthquake engineering. The success and longevity of MCEER has demonstrated that multidisciplinary research can achieve significant gains in knowledge and improvements to practice (C. Yu, University at Buffalo, personal communication, September 20, 2022).

---

contribute to corrosivity and corrosion mechanisms and translating fundamental research discoveries into practice. Multidisciplinary research in corrosion science would expose current researchers and practitioners to different ways of thinking and would provide educational opportunities to students at both the graduate and undergraduate levels. Decision making at every stage of the buried steel infrastructure life cycle can only be optimized if the knowledge from many disciplines can be effectively synthesized. This necessitates the formation of multidisciplinary teams of experts to conduct research. Such teams should include geotechnical and structural engineers, metallurgists, materials scientists, hydrologists, geochemists, geophysicists, microbiologists, and others. Topics to be explored include better ways to characterize the subsurface, combining geophysical, geochemical, hydrological, and microbiological techniques; the combined effects of different soil properties on corrosivity; and the ground response to a changing climate and its effects on corrosivity. Soong et al. (2020) predicted rapid and deep soil warming over the twenty-first century, estimating a global mean soil warming of $2.3 \pm 0.7°C$ and $4.5 \pm 1.1°C$ at 100-cm depths for two different greenhouse gas concentration trajectories. Increased temperatures will accelerate corrosion-dependent chemical reactions and will change the distribution of microorganisms and their rates of activity. Other examples of research are provided later in this chapter.

There are a variety of ways that multidisciplinary research might be facilitated. Two mechanisms are described here: formal partnerships between sectors, and multidisciplinary research centers.

Partnerships

Multidisciplinary research could be supported through the organization of formal partnerships between industry and academe, between private- and public-sector entities, and between government agencies and academic research facilities. Such partnerships between these sectors can result in the development of creative technical solutions, the translation of research concepts into practice, the development of new standards, and the strengthening of the workforce. At present, research is supported by industry or government to improve understanding of the corrosion of buried steel, but these tend to be ad hoc, focused on a specific industry sector and problem, and focused on a specific discipline or approach. The support of individual research projects remains important, but formalized programmatic-level collaborative efforts that support more than a single small-scale research project could benefit all involved. One example is Manufacturing Institutes,[1] which are industry, academe, and government partnerships created to address manufacturing issues. The institutes are funded by federal programs and through company membership dues. The inclusion of practitioners as team members in such partnerships helps to ground research in practical realities. Practitioners in both the private and public sectors benefit from the accelerated translation of fundamental research discoveries to practical applications. University researchers benefit from increased resources (e.g., funding, access to field sites, infrastructure, and data) and the opportunity to pilot innovations in the field. Students are exposed to disciplines beyond their major fields of study, have opportunities to participate in research on practical problems, and can build professional relationships to improve future job prospects. Civil engineering students with greater exposure to geophysics, for example, will better understand electrical properties of soils and the theories behind site investigation methods—a topic taught often only briefly in junior-level soil mechanics classes. Exposure to a wider array of topics helps develop professionals with a better understanding of what is possible, the significance of different types of data, and when problem solving requires additional expertise.

Multidisciplinary Research Centers

Another possible direction for facilitating multidisciplinary research is the development of multidisciplinary research centers. Several models for multidisciplinary research centers currently exist within the National Science Foundation (NSF). Once example is the NSF Science and Technology Centers: Integrative Partnerships program (NSF, 2021b), which advances interdisciplinary discovery and innovation among academic institutions, national laboratories, industrial organizations, and other public and private entities in any area of science and engineering. Another is the NSF Industry-University Cooperative Research Centers program (NSF, 2021a), which "generates breakthrough research by enabling close and sustained engagement between industry innovators, world-class academic teams and government agencies." NSF Engineering Research Centers (ERCs) are another type of support offered for multidisciplinary research. ERCs invest specifically in research that can be scaled up to technological solutions and applied in industry. These are typically 5- to 10-year programs based on a particular research theme, and they support research, teaching, and community outreach. The program has resulted in the formation of hundreds of spinoff companies, the development of new technologies and granting of hundreds of patents, and the support of thousands of undergraduate and graduate students (NSF, 2020). The U.S. Department of Defense funds the Multidisciplinary University Research Initiative (MURI) program, which "supports research teams whose research efforts intersect more than one traditional science and engineering discipline." MURI projects could be used as models for corrosion-focused centers. While the costs associated with multidisciplinary research centers can be large, benefits associated with advances in this field would also be large, given that metallic corrosion is estimated to cost the United States 3–4 percent of the U.S. gross domestic product (Koch, 2017).

COMPREHENSIVE LONGITUDINAL EXPERIMENTATION

A number of physical, chemical, and microbiological attributes of the soil environment can be related to corrosion: the type and grain size of the soil; the compaction and pore space; and the wetness, resistivity, ionic content, redox potential, oxygen concentration, pH, microbiology, and temperature. Data from statistically sound, long-term multivariate experiments that involve observations from steel buried in the subsurface make quantifying the fundamental relationships that control corrosion rates possible. Romanoff (1957) described one of the few such experimental studies on which the corrosion community relies, but conclusions drawn from those data are problematic because the soil environments in those tests were not thoroughly characterized, burial depths and

---

[1] See https://www.themanufacturinginstitute.org (accessed July 7, 2022).

exposure times varied, climate conditions were reported as averages, many soil properties were measured off-site, and the statistical design of the experiment was weak (de Arriba-Rodriguez et al., 2018). The Romanoff (1957) data need to be supplemented with better-controlled longitudinal experiments in which the same properties are repeatedly measured.

> **Recommendation 2: Coordinated groups of multidisciplinary researchers, supported through commitments from private- and public-sector organizations and agencies with interest in or responsibilities related to buried steel infrastructure, should conduct comprehensive, long-term experiments to quantify corrosion rates and mechanisms associated with multiple variables on steel buried both in controlled and in carefully characterized natural subsurface conditions.**

Comprehensive, long-term multivariate experiments (i.e., those that lead to conclusions regarding the synergistic effects of subsurface properties) need to be performed to observe the factors that contribute to corrosion and how those factors affect corrosion rates. Investigations need to allow multiple observations in the first 5 years to capture changes in corrosion rates when they are greatest, and observations 25 years or longer to capture how those rates attenuate over the service life of infrastructure. The investigations should include laboratory-based experiments with controlled initial and boundary conditions, as well as field-based experiments with extensive soil and hydrologic characterization and monitoring. Studies on buried plain carbon steel are needed, as are experiments on galvanized, aluminized, and polymer-coated steel, the long-term behaviors of which, when buried, are not well known. The latter samples (metallicized and polymer coated) would be tested using longer time frames such that degradation of the protection followed by corrosion of the underlying steel can be observed. Understanding the corrosion mechanisms likely under different conditions is also important because different mechanisms may have implications related to infrastructure decision making and management. To this end, the data collected through this effort should lead to a common validation dataset that can be used for benchmark purposes. They can then be used in a probabilistic manner to predict infrastructure reliability and to assess priorities for decision making and management.

The experiments should be designed to extract the influences of physical and chemical soil properties, soil water and gaseous phases, and soil spatial variations, and to capture the soil microbiology. Different approaches applied during the same experiments—for example, monitoring electrochemical testing and exhuming coupons for destructive testing—would benefit comparisons and integration. The experiments are intended to provide information on the performance of buried steel from an inventory of sites. Initial and emergent steel conditions will be related to climate, topography, drainage, anthropogenic activity, and details of the subsurface. An inventory of sites needs to be identified reflective of conditions that may be encountered in practice. Measurements and observations needed to characterize the subsurface include gradation, maximum particle size, Atterberg limits, pH, resistivity, salt content, organic content, redox potential, water content and degree of saturation, soil-water characterization curve, humidity, temperature, and alkalinity of the soil surrounding the metal element. Likewise, experiments should document the effects of climate change on corrosivity and corrosion of buried steel infrastructure. Descriptions of the spatial and temporal variations of these properties that may occur within the dimensions of the metal sample are also needed. Properties relating to the risk of microbially influenced corrosion (MIC) should also be quantified and monitored throughout the study. Some of the test procedures are under development and some are described within existing standards and recommended practices.

Experimental results could contribute to a reliable reference database useful to (1) identify the most relevant properties of the subsurface for corrosion rates, (2) quantify the synergistic effects of subsurface properties, (3) assess current corrosion-rate predictive models, and (4) develop corrosion models with less uncertainty in their predictive capabilities. The results will allow more strategic design of subsurface characterization and monitoring activities and inform decision making. Robust results from long-term experiments will enable designers, owners, operators, and managers to focus resources on assessing and monitoring the spatial and temporal variations of those properties with the largest impact on corrosivity and corrosion rates at a given site, thus being able to more efficiently design, construct, and manage safer and more resilient infrastructure.

Federal public agencies responsible for infrastructure management and hazard prevention and mitigation (e.g., U.S. Bureau of Reclamation and U.S. Army Corps of Engineers) could partner with public agencies with long-term funding to support longitudinal studies. Alternatively, a collection of industry groups could commit long-term funding. A good example of a long-term collaboration includes the Geosynthetic Institute (GSI,[2] originally the Geosynthetics Research Institute), which was founded in 1986 and still exists at the time of this publication (2023). The GSI is a consortium of organizations engaged in the manufacturing, design, supply, and installation of various types of geosynthetics.[3] The institute includes more than 70 members including federal and state governmental agencies, facility owners, designers, consultants, quality assurance/quality control organizations, testing laboratories, resin and additive suppliers, manufacturers, manufacturer representatives, and installation contractors who pay annual dues that sustain the institute. The GSI has developed and transferred knowledge, resources, and standards needed for the geosynthetics industry to evolve and has facilitated applications of geosynthetics in the construction industry. Longitudinal studies related to corrosion of buried steel might similarly be supported by a consortium of industry groups and state and governmental agencies.

It may be possible to leverage resources and take advantage of already established experimental sites such as those managed by various state transportation agencies to monitor corrosion.[4] Site conditions at those installations are measured and observed over time as are conditions of buried metal samples, but the installations were established for specific applications and existing data are insufficient for the types of long-term experiments recommended here. Measurement types and techniques, the properties and characteristics of the metal samples, and the subsurface conditions differ at the sites. However, existing data from these sites might inform the design of longer-term experiments, as could data found in the existing literature from laboratory experiments, and measurement of corrosion under controlled conditions. Preliminary analyses of these data would help identify data gaps. A review of the different practices would enlighten the subsurface characteristics that are most important and need to be included in plans for comprehensive and coordinated longitudinal experiments.

DATA ANALYTICS

Numerous methodologies to characterize subsurface corrosivity have been developed for specific applications (e.g., mechanically stabilized earth [MSE], soil nails, piles, culverts, pipelines; see Table 6.5), but many of those methodologies are based on a single measured property such as resistivity (NRC, 2009). Only some methods may consider the influence of multiple properties (multivariate approaches; see, e.g., Table 6.4). These multivariate approaches attempt to weight the effects of various properties on corrosivity, but the weighting factors are based primarily on judgment rather than on robust testing and modeling. None of the existing approaches incorporate all potentially relevant properties, and some include properties that may be irrelevant. Additionally, the present approaches do not describe comprehensively the synergies between subsurface properties, the value of which was recognized following presentations at the committee's workshop (e.g., presentation by Jennifer McIntosh of the University of Arizona; see Appendix B). There is a need for better understanding of the individual and combined physical phenomena that result in corrosion of buried steel, and until data from longitudinal and multivariate experiments are available, systematic examination of existing data may be useful to identify statistically important relationships among various properties and with corrosion and corrosion rates.

Recommendation 3: Researchers should use advanced data science techniques on available and new data to determine systematically the statistically important contributions of individual and combined subsurface properties to corrosivity in different subsurface environments.

---

[2] See https://geosynthetic-institute.org (accessed July 7, 2022).

[3] Geosynthetics are used in environmental, geotechnical, transportation, and hydraulic engineering. They include porous geotextiles, impermeable geomembranes, reinforcement geogrids, drainage geonets, and clay layers in/on other geosynthetics, among other materials.

[4] These include Caltrans, the North Carolina Department of Transportation, the Florida Department of Transportation, the New York State Department of Transportation, the Nebraska Department of Transportation, the South Carolina Department of Transportation, and the Wisconsin Department of Transportation. The British Columbia Ministry of Transportation and Infrastructure and the Canadian National Railway have also established sites.

There is a need to pursue modeling approaches (for both characterization and performance modeling) that are rooted in improved physical understanding of the phenomena. Longitudinal and multivariate research as described above will provide the basis for that improved understanding. However, researchers and infrastructure designers and managers could better target their resources now from systematic consideration of data from previous investigations and as new data are regularly collected. This would inform better targeted site characterization, infrastructure design, and monitoring, as well as the research needed to better understand the science behind the relationships between the physical properties and corrosion. Data analytical techniques (e.g., cluster analysis or Bayesian theory; see Chapter 8) can be applied to currently available datasets and to new data from longitudinal experiments as they become available (see Recommendation 2). These techniques should be used to investigate relationships among properties, and between properties and corrosion rates. Given enough data, machine learning techniques can be applied to identify previously unrecognized relevant subsurface properties or synergies between different subsurface properties.

In addition to understanding the relationship of properties already related to corrosion of steel, data analytical techniques may help to identify relevant subsurface properties that are not traditionally or regularly used to characterize corrosivity. For example, MIC has been attributed as a cause of failure of buried steel infrastructure (Abedi et al., 2007; California Public Utilities Commission, 2019; Kiani Khouzani et al., 2019; Sempra, 2019), but only 5 of 12 examined classification and rating schemes in Table 6.5 include measures of sulfate-reducing bacteria (SRB) or sulfides (which may indicate the activity of SRB). This lack of testing for susceptibility to MIC is likely because it is not possible to correlate numbers of particular microorganisms with the prediction or diagnosis of MIC (see Chapter 6). Furthermore, Chapter 3 describes temperature as a subsurface property that controls reaction rates, the abundance of microorganisms, and microorganism activity rates. The relationship between corrosion and temperature will be particularly relevant for retaining walls or in shallowly placed steel but less an issue for steel at greater depths. However, temperature is not considered in any of the 12 different classification and rating schemes examined in Table 6.5.

Although Chapter 2 describes numerous experimental studies that have measured corrosion on buried steel across a number of climates, these studies, to the committee's knowledge, have not been systematically "mined" for data to assess the relationship between temperature and corrosion and corrosion rates, or for their relationships to MIC. Given changing climates and the higher likelihood of many geographical locations to experience temperature extremes, such information could be valuable. Similarly, Little et al. (2020) reported that more than 2,000 papers on MIC had been published in the previous 25 years that describe anecdotal failures associated with MIC as well as laboratory and field testing conducted under varying conditions (e.g., natural microbial populations versus pure cultures of a single microorganism; natural electrolytes such as seawater, estuarine water versus enriched laboratory media that did not approximate any natural electrolyte; and varying temperatures, exposure durations, and metallurgies). The result is a collection of independent observations that have not produced a predictive capability for any material–microorganism–metal substratum. The collective data for MIC are extensive and could be systematically examined for relationships directly related to MIC (e.g., assimilable nutrients, relationships between electron donors or electron acceptors and aggressive or inhibiting anions). The results will provide an integrated approach to predict MIC that is based on the total environment and not the identification of specific putative microorganisms.

Improved estimates of corrosion rates will result from analytical approaches that (1) consider all relevant subsurface properties, (2) apply data-driven weighting factors to relevant subsurface properties, and (3) calculate the synergies between the relevant subsurface properties. However, data analytical approaches alone cannot improve basic fundamental understanding of the underlying physical processing influences, and advanced predictive modeling will only be improved with improvements in physics-based modeling. The results of successful experiments will direct the future development of new characterization methodologies and ultimately inform enhanced ability to anticipate failures, estimate remaining service life for existing facilities, and incorporate efficient corrosion management practices into designs.

DECISION SUPPORT SYSTEMS

The fundamental mechanics of corrosion are the same for the pipeline and geo-civil industries, as is the need to make decisions that prioritize actions and investments where they are most impactful. However, decisions related to site characterization are often made in an ad hoc manner or following specific industry standards. For example, the geo-civil industries conduct site investigations to determine the mechanical and hydraulic properties needed for design and analysis of foundation systems, global stability, drainage, and problems that involve transport, but few site characterization protocols guide proper data collection for characterizing corrosivity and corrosion modeling. The same can be said for management decisions concerning previously buried steel assets. Decisions based solely on individual past practices or industry-specific standards do not benefit from the experience of other industries or new innovations.

A DSS is a tool that guides decision makers through alternatives. Global positioning system–based navigation systems are examples of DSSs in common use. The systems suggest routes to a requested location based on input from the user (e.g., the desire to avoid certain types of roads, minimize travel times, or maximize fuel efficiency) and incorporate information about historical travel times, road closures, traffic, and construction into suggested alternatives with information about expected travel times (i.e., outcomes). The systems are sophisticated enough to refine the route based on real-time changing conditions (e.g., newly reported accidents). A DSS can be as simple as a two-dimensional flowchart guiding choices between binary options (see Figures 6.6 and 6.7), or it can be a complex digital system connected to multiple input databases that leads a decision maker through numerous options. GeoTechTools[5] is a DSS developed by the Strategic Highway Research Program of the National Academies of Sciences, Engineering, and Medicine, deployed by the Federal Highway Administration. It is now hosted by the Geo-Institute of the American Society of Civil Engineers (ASCE) and is used in the engineering community to support ground improvement decisions.

Whether a simple flowchart or sophisticated computer algorithm, DSSs for engineers are designed to categorize and rank alternatives based on data, models, design standards, and engineering judgment, and can be used to define the uncertainties associated with specific techniques and methods. For example, to compute corrosion rates from linear polarization resistance (LPR) data, one must assume a certain rate-controlling process (that the corrosion reaction is activation controlled), fit a line to the plot of overpotential versus corrosion current measurements, and determine values for Tafel constants, the surface area, density, equivalent molecular weight, and valance of the metal. These quantities are rated to metal type and, for a galvanized element, whether zinc or base steel exposure at the surface may be uncertain. The element under test must be electrically isolated such that the surface area involved in the metal loss is known. These uncertainties can be revealed or reduced by scrutinizing various aspects of the data (e.g., the measured corrosion potential can indicate whether zinc or steel is exposed at the surface, and soil resistance is indicative of the surface area of the element included with the measurement). Performing an LPR test is a simple matter, but discerning if the data are good requires knowledge and experience. A robust DSS will help to reduce uncertainty by formalizing standard practices and present logical and reproducible sequences of decisions based on existing data. They could be particularly helpful in decision areas that rely on large volumes of data combined with predictive modeling.

Simplified and empirical methods for modeling metal loss, corrosion rates, and performance of protection systems have limitations and are only reliable for particular sets of conditions. Without a common database of reported case histories of failures, it is difficult to assess past performance of given protection systems in given environments and validate models. Deciding which model is best applied for choosing a protection system or amount of steel to add to the infrastructure to account for steel loss for a particular site, depth of burial, specific design details, and other factors could be assisted with the use of a DSS. In theory, a single DSS could support decisions through all stages of infrastructure design life, but development and implementation of separate DSSs for site characterization and monitoring might be more practical. A site characterization DSS is needed to guide the formulation of site characterization plans that capture the individual and combined subsurface characteristics that affect corrosivity as well as the important aspects of lateral, vertical, and temporal variabilities and the uncertainties

---

[5] See geotechtools.org (accessed July 7, 2022).

associated with those variabilities. Such a DSS would be an improvement over current standard characterization practices in all industries managing corrosion.

Recommendation 4: Standards-setting bodies should collaborate with state and federal agencies, industry groups, and academe to create and maintain two decision support systems (DSSs):

(1) a DSS that guides site characterization and allows selection from among a comprehensive set of characterization tests that are appropriate for temporally and spatially variable surface and subsurface conditions; and

(2) a DSS that uses risk-informed decision making to guide corrosion management practices.

The simple DSS presented in Figures 6.6 and 6.7 guides the choice of tests for pH, salt content, and resistivity depending on the gradation of the sample and correlation of those properties given observed conditions. That DSS only includes standard tests for a few properties, and only those developed by AASHTO and the Texas Department of Transportation. Test standards have been developed by other state transportation agencies and promulgated by various agencies or industry groups such as the U.S. Environmental Protection Agency, ASTM International, the American Water Works Association, and the Soil Science Society of America. Many regionally or industry-developed test standards could be useful but are likely unknown outside the region or industry for which they were developed. The DSS in Figures 6.6 and 6.7 does not provide information regarding how uncertainties in the results might influence decisions under various conditions. A common, more comprehensive characterization framework and DSS is needed that informs decisions related to subsurface characterization appropriate for multiple combinations of subsurface properties. A DSS for practitioners should outline the minimum information needed to design a site characterization program, and it should provide guidance regarding preliminary field and laboratory tests and spatial sampling frequencies needed based on the natural setting of the site, land use, infrastructure life cycle, surface and groundwater hydrology, and atmospheric conditions. The framework and DSS should then help guide decisions regarding additional characterization necessary to reduce uncertainties to acceptable levels for modeling.

To make its development a more practical exercise, the DSS could be developed in stages, with earlier versions of the DSS guiding decisions about tests needed to characterize the subsurface properties most utilized to model corrosivity (e.g., moisture content, resistivity, pH, chlorides, and sulfates; see Table 6.5). Later versions can be expanded to include guidance regarding promising but less commonly measured properties (sulfides, microbial-related properties, and redox potential). The system should include guidance regarding both laboratory- and field-based methods and should distinguish which laboratory tests are intended to replicate field conditions versus those that do not. Uncertainties associated with each method need to be identified, examples of method application should be provided, and the properties under which the test is accurate should be indicated. This includes guidance regarding how to understand errors in modeling given the disparities between the scales of measurements and the scales at which corrosion is initiated on the steel surface. Guidance regarding how to account for spatial and temporal variation in the surface and subsurface is an important aspect of the DSS that also needs to be incorporated into the system. As the community gains understanding regarding the multivariate controls on corrosivity (see Recommendation 2), the DSS should then be expanded to include guidance informed by those controls. Likewise, new test methods based on a multidisciplinary understanding of the subsurface should be incorporated into the DSS as their results are understood.

As the characterization DSS is developed, a second DSS based on risk-informed decision making (i.e., informed by the likelihood and severity of negative impacts) should be developed to inform management actions and investments. This second DSS should be developed and maintained in parallel or in concert with the characterization DSS so that it uses outputs from the characterization DSS (including present and future uncertainties about the environment in which the steel is buried) as part of its input. The DSS could be applied to risk-based decision making associated with both new infrastructure design and the modeling and monitoring of existing infrastructure. The DSS could provide screening and selection criteria for constructed earth that mitigate steel corrosion (i.e., that control pH, resistivity, salt content, and organics content).

To advance appropriate and useful interindustry utilization of monitoring techniques, standards-setting bodies should work with state and federal agencies and industries to develop a common corrosion management practices DSS. This will require coordinated input, planning, and action of all agencies and organizations with interest in or responsibilities associated with corrosion of buried steel. These groups will need to develop a framework that is able to tie available multivariate information and guide prioritization of actions using risk-informed approaches. The DSS would be based on comprehensive information regarding protection methods (physical barriers and CP systems), the type of environments for which they are well suited, the environments in which they fail, and the combinations of factors within those environments that resulted in the failures. Terminology and descriptions used in the DSS would be drawn from a common lexicon (see Recommendation 1). The system would indicate what coatings are susceptible to microbial degradation (see Chapter 4) and would allow practitioners to input conditions under which the protective system failed. It should be interactive and guide selection of protective design systems by practitioners, given selected inputs (e.g., electrical continuity, depth to infrastructure). The DSS should guide decisions regarding monitoring techniques depending on a number of inputs, including the variable to be monitored (coating defects, corrosion rate, CP effectiveness), the depth and dimensions of the infrastructure, and the risk associated with failure. This DSS would serve the broader community, including smaller organizations (e.g., small water utilities) that may not have expertise in all areas, and would assist the geo-civil industries in improving asset management (see Chapter 7).

INDIRECT OBSERVATION AND OPPORTUNISTIC DATA COLLECTION

Corrosion of buried steel can be monitored directly and indirectly through destructive and electrochemical tests on the infrastructure itself, such as those described in Chapter 7. Alternatively, if the subsurface environment is thoroughly understood and defined, it is theoretically possible to predict which corrosion mechanisms are likely to occur at what rates. While the risks associated with the failure of some infrastructure may support implementation of robust subsurface and infrastructure monitoring programs, such programs may not always be economically feasible. For example, it is not feasible to monitor subsurface properties across hundreds of kilometers in oil and gas projects. It also is not economically feasible to repeatedly expose geo-civil steel infrastructure for direct monitoring. Finally, it is not feasible (or, at present, possible) to quantify all the properties relevant to corrosivity for entire infrastructure systems, and certainly not continuously for the duration of the infrastructure life cycle. However, surface monitoring (indirect observations) and opportunistic data collection could inform where localized site-specific monitoring is warranted. Those data could be used to build a database to inform future research and infrastructure-related decision making.

Recommendation 5: Private- and public-sector infrastructure owners should monitor the land surface for changes that could alter subsurface corrosivity and determine whether localized monitoring of subsurface properties is warranted to maintain infrastructure performance and safety. Surface changes to be monitored include but are not limited to changes in land, land use, and atmospheric conditions that affect surface and groundwater flow, and any asset management decisions by colocated infrastructure managers that might affect subsurface hydrology, geochemistry, microbiology, or the production of stray currents.

Ideally, research and development could lead to future capacity to directly measure corrosivity, corrosion, and corrosion rates. At present, infrastructure managers must rely on indirect measurements to estimate corrosivity and corrosion rates. Because changes on the land surface can affect surface and groundwater flow, permeability, soil saturation, soil and water chemistries, subsurface temperatures, and other characteristics that affect corrosivity, monitoring changes on the surface provides a cost-effective early indicator of possible detrimental changes in the subsurface. Noting changes on the land surface or in management practices would be instructive of when and where more direct measurements are appropriate.

Surface monitoring should include monitoring changes in

- land use, including upgradient land use, installation of pavements, large foundations, or other underground structures, and installation of surface, subsurface, or aerial transmission or pipelines that may produce stray currents (e.g., including from CP);
- land cover, such as transition from rural to urban, and changes in vegetation, such as those that may indicate moisture accumulation;
- installation of upgradient power plants, mining operations, or waste disposal operations that could affect groundwater and soil geochemistries;
- surface water- or groundwater flow or retention, changes at the surface that would alter surface-water flow or retention, or the appearance or disappearance of springs;
- infrastructure conditions such as those that result in the intentional or unintentional release of fluids or change in subsurface temperatures;
- infrastructure or land management decisions such as use of or change in deicing salts on pavements or the application of fertilizers that could leach into the subsurface and increase electrochemical potential or encourage the growth of microbes that influence corrosion;
- changes in construction practices; and
- climate and atmospheric conditions (seasonal and global) that result in changes in temperature and precipitation that in turn affect subsurface temperature, groundwater and the groundwater table, degree of saturation, and saltwater intrusion and that may alter the magnitude or frequency of extreme events.

Based on surface conditions, it would then be feasible to monitor key sites (e.g., a mining operation upgradient from a retaining wall) or to establish site-specific and localized monitoring based on identified corrosion mechanisms, environments, or infrastructural components. For example, if it was learned that MIC is possible at a site, a localized monitoring program such as was established for the new (2021) Frederick Douglass Memorial Bridge on the Anacostia River (R. Poston, Pivot Engineers, personal communication, May 26, 2022) could determine whether corrosive conditions persist.

Monitoring some surface changes can be relatively inexpensive and could even be accomplished from one's desktop with few computational resources—for example, tracking information about precipitation, surface temperatures, land use changes, traffic pattern changes, and changes in topography that might be publicly available from the Internet. Other monitoring could be accomplished by installing or retrofitting infrastructure with sensors—for example, fitting sections of pavement to monitor the effects of seasonal rainfall in the first few feet of the subsurface. Moisture sensors, resistivity meters, in situ pH sensing with specific ranges, and lysimeters (e.g., those installed at the base of landfills to collect in situ pore water) are available or are in various stages of development and could be incorporated into "smart structures" to monitor for corrosivity (Drumm et al., 2006; Genc et al., 2019; Hinshaw and Northrup, 1991; Lee et al., 2009; Liang et al., 2006; Loi et al., 1992; McCartney and Khosravi, 2013; Taamneh and Liang, 2010). However, understanding the relevance of all this information requires longitudinal research and data not only of the type described in Recommendation 3 but also on topics such as long-term fate and transport and unsaturated soil mechanics (i.e., percent saturation and hydraulic conductivity). Tracking, recording, and recognizing the significance of these and other types of data across infrastructure types will require a systems management approach. A manager of subsurface steel infrastructure might, for example, benefit from knowing that another infrastructure manager changed deicing salts applied to nearby pavement and understanding how the change in salts might change soil and water chemistries and therefore corrosivity.

Recommendation 6: Private- and public-sector infrastructure owners should capitalize on opportunities to record properties of the subsurface and steel in a standardized way when infrastructure needs to be maintained, decommissioned, or replaced.

There are numerous opportunities for opportunistic observation, inspection, and data collection in the geo-civil industries. These opportunities exist when infrastructure reaches the end of a service life or is undergoing improvements and is partially or completely decommissioned or replaced. At these times, standardized protocols to collect data regarding subsurface properties and infrastructure corrosion should be implemented. Developing and implementing data-gathering protocols at such times could greatly increase data availability. Additionally, if it is noticed that a type of steel component is prone to failure in a particular structure or in certain environmental conditions, monitoring that component type or those conditions might be warranted. For example, during expansion projects when MSE walls are partially deconstructed, samples of galvanized steel reinforcements may be observed as they are exhumed and measurements made to describe the conditions and metal losses. There may be no expectation of performance-threatening corrosion, but if found, retrofit or rehabilitation of the MSE walls may be necessary (Nicks et al., 2015, 2017). Because excavation is costly, infrastructure owners should take advantage of unexpected opportunities to monitor steel, collect subsurface information, and track infrastructure and subsurface changes. Data from fortuitous monitoring opportunities should be systematically saved to inform longitudinal research, DSS, and future decisions for that infrastructure and for buried steel infrastructure more generally.

A DATA CLEARINGHOUSE

There is a general lack of access to data with which to build and strengthen corrosion-related models. As discussed in Chapter 2, engineers addressing corrosion of buried steel rely on limited data published in the middle of the past century (Romanoff, 1957) to predict corrosivity and inform corrosion models. Those data are useful, but their utility is limited. In some cases, other data may exist, but either there is no platform from which those data may be accessed, or the data may be proprietary and not publicly accessible. Researchers; infrastructure designers, owners, and managers; and steel and steel protection manufacturers could benefit from a public-domain data clearinghouse from which standardized data from multiple industries can be queried and combined to better inform empirically based corrosion rate modeling and corrosion prediction capabilities. The ability to mine data and information would inform corrosion-related experimental designs (see the above recommendations) and inform modeling and analysis used for infrastructure-related design and decision making.

Recommendation 7: Industry groups, public-sector agencies with responsibilities related to buried steel infrastructure, and research organizations should coordinate to establish a public-domain data clearinghouse organized around consistent data-recording standards and a common lexicon for secure sharing of data related to the corrosion of buried steel including data on soil environment, corrosion potential and rates, and corrosion monitoring data.

Public-domain cyber infrastructure platforms for engineering data exist in which standardized data are deposited by researchers, infrastructure owners, and industry. These were created to allow access to and use of the data. Some examples are the Natural Hazards Engineering Research Infrastructure DesignSafe-CI, funded by NSF (NHERI, 2022); and the Collaborative Reporting for Safer Structures US,[6] which is a database of safety information for structural engineers, supported by the Structural Engineering Institute of ASCE. The International Stormwater Best Management Practices Database (BMPDB[7]; see Box 9.2), established in 1996, is an example repository designed to archive field data gathered from the design and performance of stormwater best management practices. Like buried steel, the performance of stormwater systems is influenced by numerous variables, so the BMPDB is a good example framework for collecting data that might inform a corrosion-related database.

Schema for consistent data terminology intended to facilitate data transfer have also been developed, for example, by the Data Interchange for Geotechnical and Geoenvironmental Specialists (DIGGS),[8] which is a data transfer protocol supported by the ASCE Geo-Institute. DIGGS follows a protocol developed by the Association of Geotechnical and Geoenvironmental Specialists (AGS) and localized for a given region (e.g., New Zealand Geotechnical Society Inc., 2017). The geotechnical data interchange protects and may enhance data integrity and

---

[6] See https://www.cross-safety.org/us (accessed April 5, 2022).

[7] See https://bmpdatabase.org (access July 8, 2022).

[8] See https://www.geoinstitute.org/special-projects/diggs (accessed July 8, 2022).

BOX 9.2
The International Stormwater Best Management Practices Database

The International Stormwater Best Management Practices (BMP) Database[a] project began in 1996 under a cooperative agreement between the American Society of Civil Engineers (ASCE) and the U.S. Environmental Protection Agency. The project was originally intended to inform the design, selection, and performance of urban stormwater BMPs, but later was augmented to include information about agricultural runoff, treatment, and management (Clary et al., 2020) and information about stream restoration techniques.[b] In 2004, the project transitioned to a broader coalition of partners now led by the Water Research Foundation, including the Environmental and Water Resources Institute of ASCE, the Federal Highway Administration, and other sponsors over the years. Continued population of the database and assessment of its data will ultimately lead to a better understanding of factors influencing BMP performance and help to promote improvements in BMP design, selection, and implementation.

---

[a] See https://bmpdatabase.org (accessed July 8, 2022).
[b] See https://bmpdatabase.org/stream-restoration-database (accessed July 8, 2022).

---

allows transfer of data between stakeholders (e.g., project owners, engineers, and technicians) in a data structure that can be fused and fed seamlessly into processing applications, such as data visualization. Although difficult to implement in the United States because of the diversity of data collection methodologies used by 50 different states, DIGGS is being adopted elsewhere, such as in New Zealand and the United Kingdom. Some platforms were originally developed for research purposes but are growing to serve industry and to move state-of-the-art technologies and knowledge to practice.

Developing a data clearinghouse is a long-term investment for the field of underground corrosion. Other technical communities, such as the biomedical research community, who are more mature in their efforts to create data platforms and are committed to the concept of making data findable, accessible, interoperable, and reusable (known as FAIR data) should be considered (Wilkinson et al., 2016). Experts from appropriate engineering and scientific disciplines, as well as the relevant data scientists, software engineers, information technology specialists, and data curators, should be convened to identify the mission and goals of a clearinghouse. Decisions regarding the type of data resources to be made available, characteristics of typical data contributors and users, the potential value of the data now and in the future, the infrastructure and personnel necessary to manage the clearinghouse over the short and long terms, the major cost drivers, and even associated lifetime costs of curating and managing data over the short and long terms need to be considered (e.g., NASEM, 2020).

The results of this effort would be a searchable repository of observations and measurements that can describe the effects of subsurface properties and characteristics on steel and variations of these quantities on the durability, performance, and corrosion rates of buried steel in a variety of applications. The platform will provide researchers the necessary data to accelerate fundamental research, which will in turn advance the state of the art and of practice in industry. Given proprietary or security and vulnerability concerns, data could be identified uniquely and without relation to specific location or infrastructure, which may encourage industry to include its data for the benefit of the entire technical community. Ultimately, as more data are deposited and made available, advanced data analytical techniques including artificial intelligence and machine learning may be used to mine and analyze the data and enable a more holistic understanding of the environmental contributors to corrosivity and corrosion rates. With increased availability of standardized, multidisciplinary, and high-quality data collected from well-documented sites, engineering practitioners could investigate and better understand the contributions of combined subsurface properties to corrosivity and corrosion rates in a given type of environment. Future site characterization investigations can be designed more effectively, infrastructure design and management can be more efficient, and monitoring programs can target environments and conditions shown to be problematic for certain types of infrastructure.

Whereas designing, determining the governance for, and establishing a clearinghouse may take years, steps can be taken in the shorter term to identify existing data that should be included in the clearinghouse. These might be the data types such as those included in Table 6.5 of this report, and consideration should be given as to how they should be reported and what other data might be included. As such decisions are made, it will be necessary to develop common data-recording standards (including for metadata) for future data collections.

CONCLUDING THOUGHTS AND MOVING FORWARD

Corrosion is a trillion-dollar problem (Koch, 2017) that requires knowledge from a variety of disciplines—from microbiology to metallurgy and various fields of engineering—and it affects a wide array of industries. After steel is buried in the ground, it cannot be easily observed without expensive excavations or complex monitoring programs. The implementation of the recommendations in this report will require extensive support and resources from and coordination and cooperation among industry sectors and experts from the public and private sectors and academe. The study committee was not charged with considering what agencies and organizations should be responsible for leading or funding implementation of the recommendations, nor was the committee constituted to deliberate the economic and policy considerations that should inform such recommendations. However, coordinated effort among a variety and range of organizations will be necessary. Implementation of the recommendations may require, for example, cross-disciplinary and cross-sector workshops to identify interested and affected parties, needs and priorities, commonalities and differences in the understanding of key concepts, and existing resources, and to organize and coordinate processes and responses. Broadening and deepening engagement with a broad base of interested agencies and organizations at the earliest stages of implementation will result in better identification of issues, resources, and solutions. However difficult implementation will be, it will result in improved understanding and communication and will allow for model validation, more realistic prediction of corrosion, and more efficient and cost-effective engineering and infrastructure management.

The significantly different approaches of the geo-civil and oil and gas pipeline industries have resulted in intellectual and practical silos. This situation is made worse because researchers and practitioners rarely move between the pipeline and geo-civil industries. These silos lead to parallel—or even competing—research initiatives and foci. The recommendations in this report are intended to help break down the silos and encourage the multidisciplinary and multiscale considerations from atomistic to global climate change. Significantly improving understanding of corrosion mechanisms and rates for buried steel will require a multidisciplinary approach with an inclusive vocabulary that is easily translated between disciplines. As comprehensive long-term multivariate experiments are conducted and observational data are collected, reliable and accessible databases can be established. Data support systems for site characterization program design and risk-informed decision making can be developed. Advanced data analytics should be applied using the current dataset, but more complicated methods such as machine learning will benefit from this robust, collective database.

The committee envisions that industry groups will work through their memberships to develop research needs statements and calls for research proposals based on the needs identified in this report. Ultimately, the recommendations will lead to an understanding of corrosion that will improve the ability to predict corrosion mechanisms and rates, protect against that corrosion, and monitor and model to predict performance more accurately. It is the committee's expectation that the recommendations will also lead to movement in industry-specific practices, where geo-civil works will adopt protection and monitoring approaches where they make sense, and pipeline industries might adopt characterization approaches used in geo-civil practices. Changes that result from the implementation of the recommendations could reduce the costs of maintaining safety and the environment, and for operation and preservation of buried steel infrastructure.

References

AASHTO (American Association of State Highway and Transportation Officials). 1990. *Guidelines for pavement management systems*. Washington, DC: AASHTO.

AASHTO. 2002. *Standard specifications for highway bridges*. Washington, DC: AASHTO.

AASHTO. 2020. *LRFD bridge design specifications*, 9th ed. Washington, DC: AASHTO.

AASHTO R 27. 2001. *Standard practice for assessment of corrosion of steel piling for non-marine applications*. Washington, DC: AASHTO.

AASHTO T 265-15. 2019. *Laboratory determination of moisture content of soils*, 41st ed. Standard specifications for transportation materials and methods of sampling and testing and provisional standards. Washington DC: AASHTO

AASHTO T 267-86. 2018. *Determination of organic content in soils by loss on ignition*, 41st ed. Standard specifications for transportation materials and methods of sampling and testing and provisional standards. Washington, DC: AASHTO.

AASHTO T 288-12. 2016. *Determining minimum laboratory soil resistivity*, 41st ed. Standard specifications for transportation materials and methods of sampling and testing and provisional standards. Washington, DC: AASHTO.

AASHTO T 289-91. 2018. *Determining pH of soil for use in corrosion testing*, 41st ed. Standard specifications for transportation materials and methods of sampling and testing and provisional standards. Washington, DC: AASHTO.

AASHTO T 290-95. 2020. *Determining water-soluble sulfate ion content in soil*, 41st ed. Standard specifications for transportation materials and methods of sampling and testing and provisional standards. Washington, DC: AASHTO.

AASHTO T 291-94. 2018. *Determining water-soluble chlorine ion content in soil*, 41st ed. Standard specifications for transportation materials and methods of sampling and testing and provisional standards. Washington, DC: AASHTO.

Abedi, S. S., A. Abdolmaleki, and N. Adibi. 2007. Failure analysis of SCC and SRB induced cracking of a transmission oil products pipeline. *Engineering Failure Analysis* 14(1):250-261.

Adkins, G., and N. Rutkowski. 1998. Field and laboratory resistivity testing of mechanically stabilized earth systems (MSES) backfill material in 1996 and 1997. In *48th Highway Geology Symposium Proceedings and Fieldtrip Excursion Guide*. Albany, NY: New York State Department of Transportation Geotechnical Engineering Bureau.

AGA (American Galvanizers Association). 2010. *Performance of hot-dip galvanized steel products: In the atmosphere, soil, water, concrete, and more*. Centennial, CO: AGA.

Agarry, S., and K. Salam. 2016. Modelling the kinetics of microbiologically influenced corrosion of mild steel in soil environments. *Thamassat International Journal of Science and Technology* 21(4):44-65.

Akhoondan, M., and A. A. Sagüés. 2013. Corrosion mechanism of aluminized steel in limestone backfill. *Corrosion* 69(12):1147-1157.

Akhoondan, M., A. A. Sagüés, and L. J. Cáseres. 2008. *Corrosion assessment of mechanically formed aluminized steel*. Paper presented at CORROSION 2008, New Orleans, Louisiana. Paper No. NACE-08396.

Al-Thawadi, S. M. 2011. Ureolytic bacteria and calcium carbonate formation as a mechanism of strength enhancement of sand. *Journal of Advanced Science and Engineering Research* 1(1):98-114.

AMPP (Association for Materials Protection and Performance) TM0497-2018-SG. 2018. *Measurement techniques related to criteria for cathodic protection on underground or submerged metallic piping systems.* Houston, TX: AMPP.

Anderko, A. M., R. D. Young, and P. McKenzie. 2001. Computation of rates of general corrosion using electrochemical and thermodynamic models. *Corrosion* 57:202-213.

Angst, U. M. 2019. A critical review of the science and engineering of cathodic protection of steel in soil and concrete. *Corrosion* 75(12):1420-1433.

ANSI/AWWA (American National Standards Institute and American Water Works Association) C105/A21.5. 2018. *American national standard for polyethylene encasement for ductile-iron pipe systems.* American National Standards Institute and American Water Works Association.

Ansuini, F. J., and J. R. Dimond. 1994. Factors affecting the accuracy of reference electrodes *Materials Performance* 33(1):14-17.

Arashi, M., A. M. E. Saleh, and B. G. Kibria. 2019. *Theory of ridge regression estimation with applications.* Hoboken, NJ: John Wiley & Sons.

Arciniega, J. L., D. Cabrera, T. Svede, S. Rocha, J. Garibay, A. Bronson, W. S. Walker, and S. Nazarian. 2021. *Implementation of new specification requirements for coarse backfill materials for mechanically stabilized earth (MSE) walls.* Research Report FHWA/TX-18/5-6359-01-1. Springfield, VA: National Technical Information Service.

Asadi, Z. S., and R. E. Melchers. 2017. Extreme value statistics for pitting corrosion of old underground cast iron pipes. *Reliability Engineering and System Safety* 162:64-71.

ASME (American Society of Mechanical Engineers) B31G-2012 (R20170). 2017. *Manual for determining the remaining strength of corroded pipelines.* New York: American Society of Mechanical Engineers.

ASTM A123/A123M-17. 2017. *Standard specification for zinc (hot-dip galvanized) coatings on iron and steel products.* Vol. 01.06. West Conshohocken, PA: ASTM International.

ASTM A153/A153M-16a. 2016. *Standard specification for zinc coating (hot-dip) on iron and steel hardware.* Vol. 01.06. West Conshohocken, PA: ASTM International.

ASTM A929/A929M-18. 2021. *Standard specification for steel sheet, metallic-coated by the hot-dip process for corrugated steel pipe.* Vol. 01.06. West Conshohocken, PA: ASTM International.

ASTM B633-19. 2019. *Standard specification for electrodeposited coatings of zinc on iron and steel.* Vol. 02.05. West Conshohocken, PA: ASTM International.

ASTM C876-15. 2016. *Standard test method for corrosion potentials of uncoated reinforcing steel in concrete.* Vols. 03.02, 04.02. West Conshohocken, PA: ASTM International.

ASTM C1580-20. 2021. *Standard test method for water-soluble sulfate in soil.* Vol. 04.02. West Conshohocken, PA: ASTM International.

ASTM D422-63. 2016. *Standard test method for particle-size analysis of soils* (withdrawn 2016). West Conshohocken, PA: ASTM International.

ASTM D888-18. 2018. *Standard test methods for dissolved oxygen in water.* Vol. 11.01. West Conshohocken, PA: ASTM International.

ASTM D1067-16. 2016. *Standard test methods for acidity or alkalinity of water.* Vol. 11.01. West Conshohocken, PA: ASTM International.

ASTM D2216-19. 2019. *Standard test methods for laboratory determination of water (moisture) content of soil and rock by mass.* West Conshohocken, PA: ASTM International.

ASTM D2487-17e1. 2020. *Standard practice for classification of soils for engineering purposes* (Unified Soil Classification System). Vol. 04.08. West Conshohocken, PA: ASTM International.

ASTM D3875-15. 2017. *Standard test method for alkalinity in brackish water, seawater, and brines.* Vol. 11.02. West Conshohocken, PA: ASTM International.

ASTM D4219-08. 2017. *Standard test method for unconfined compressive strength index of chemical- grouted soils* (withdrawn 2017). West Conshohocken, PA: ASTM International.

ASTM D4318-17e1. 2018. *Standard test methods for liquid limit, plastic limit, and plasticity index of soils.* Vol. 04.08. West Conshohocken, PA: ASTM International.

ASTM D4327-17. 2019. *Standard test method for anions in water by suppressed ion chromatography.* Vol. 11.01. West Conshohocken, PA: ASTM International.

ASTM D4542-15. 2016. *Standard test methods for pore water extraction and determination of the soluble salt content of soils by refractometer.* Vol. 04.08. West Conshohocken, PA: ASTM International.

ASTM D4658-15. 2017. *Standard test method for sulfide ion in water.* Vol. 11.01. West Conshohocken, PA: ASTM International.

ASTM D4972-19. 2019. *Standard test methods for pH of soils*. Vol. 04.08. West Conshohocken, PA: ASTM International.

ASTM D6386-16. 2016. *Standard practice for preparation of zinc (hot-dip galvanized) coated iron and steel product and hardware surfaces for painting*. Vol. 06.02. West Conshohocken, PA: ASTM International.

ASTM D7396-14(2020). 2020. *Standard guide for preparation of new, continuous zinc-coated (galvanized) steel surfaces for painting*. Vol. 06.02. West Conshohocken, PA: ASTM International.

ASTM G51-18. 2021. *Standard test method for measuring pH of soil for use in corrosion testing*. Vol. 03.02. West Conshohocken, PA: ASTM International.

ASTM G57-20. 2020. *Standard test method for measurement of soil resistivity using the Wenner four-electrode method*. Vol. 03.02. West Conshohocken, PA: ASTM International.

ASTM G59-97. 2020. *Standard test method for conducting potentiodynamic polarization resistance measurements*. Vol. 03.02. West Conshohocken, PA: ASTM International.

ASTM G187-18. 2018. *Standard test method for measurement of soil resistivity using the two-electrode soil box method*. Vol. 03.02. West Conshohocken, PA: ASTM International.

ASTM G200-20. 2020. *Standard test method for measurement of oxidation-reduction potential (ORP) of soil*. Vol 03.02. West Conshohocken, PA: ASTM International.

ASTM WK24621. 2015. *Measurement of coarse aggregate resistivity using the two-electrode soil box*, draft document, 08-28-2015, for consideration by ASTM subcommittee c09.20. West Conshohocken, PA: ASTM International.

Atha, D. J., and M. R. Jahanshahi. 2018. Evaluation of deep learning approaches based on convolutional neural networks for corrosion detection. *Structural Health Monitoring* 17(5):1110-1128.

Ault, J. P., and J. A. Ellor. 2000. *Durability analysis of aluminized Type 2 corrugated metal pipe*. FHWA-RD-97-140. McLean, VA: Federal Highway Administration.

AWS (American Welding Society) C2.23M/C2.23:2018. 2018. *Specification for the application of thermal spray coatings (metallizing) of aluminum, zinc, and their alloys and composites for the corrosion protection of steel*. Miami, FL: American Welding Society.

AWWA (American Water Works Association) M77. 2019. *Condition assessment of water mains*. Denver, CO: American Water Works Association.

Aziz, P. 1956. Application of the statistical theory of extreme values to the analysis of maximum pit depth data for aluminum. *Corrosion* 12(10):35-46.

Banciu, H. L. 2013. Diversity of endolithic prokaryotes living in stone monuments. *Studia Universitatis Babes-Bolyai, Biologia* 58(1):99-109.

Barlo, T., and W. Berry. 1984. An assessment of the current criteria for cathodic protection of buried steel pipelines. *Materials Performance* 23(9).

Barros, N. 2021. Thermodynamics of soil microbial metabolism: Applications and functions. *Applied Sciences* 11(11):4962.

Bastick, M., and J.-M. Jailloux. 1992. Twenty-five years of corrosion control in reinforced earth structures. In *Earth reinforcement practice*, edited by H. Ochiai, S., Hayashi, and J. Otani. Rotterdam, Netherlands: A. A. Bakema. Pp. 17-22.

Baxter, D. 1996. Do all rockbolts rust? Can QA help? But does it matter? *Breaking new ground: IX Australian Tunnelling Conference Proceedings, Sydney, Australia, 27–29 August*. Institution of Engineers, Australia.

Bazán, F. A. V., and A. T. Beck. 2013. Stochastic process corrosion growth models for pipeline reliability. *Corrosion Science* 74:50-58.

BEASY. 2022. *Galvanic corrosion simulation software*. https://www.beasy.com/galvanic-corrosion/software-solutions.html (accessed May 20, 2022).

Beavers, J. A. 2001. Cathodic protection—how it works. In *Peabody's control of pipeline corrosion*, edited by R. L. Bianchetti. Houston, TX: NACE International. Pp. 21-26.

Beavers, J. A. 2014. 2013 Frank Newman Speller Award lecture: Integrity management of natural gas and petroleum pipelines subject to stress corrosion cracking. *Corrosion* 70(1):3-18.

Beavers, J. A., and C. L. Durr. 1998. *NCHRP Report 408: Corrosion of steel piling in nonmarine applications*. Washington, DC: Transportation Research Board and National Academy Press.

Berg, R. R., N. C. Samtani, and B. R. Christopher. 2009. *Design of mechanically stabilized earth walls and reinforced soil slopes*, volume II. Federal Highway Administration.

Bianchetti, R. L. 2001. Survey methods and evaluation techniques. In *Peabody's control of pipeline corrosion*, edited by R. Bianchetti. Houston, TX: NACE International. Pp. 75-77.

Binley, A., and L. Slater. 2020. *Resistivity and induced polarization: Theory and applications to the near-surface earth*. Cambridge, UK: Cambridge University Press.

Bintrim, S. B., T. J. Donohue, J. Handelsman, G. P. Roberts, and R. M. Goodman. 1997. Molecular phylogeny of archaea from soil. *Proceedings of the National Academy of Sciences of the United States of America* 94(1):277-282.

Birbilis, N., M. K. Cavanaugh, A. D. Sudholz, S. M. Zhu, M. A. Easton, and M. A. Gibson. 2011. A combined neural network and mechanistic approach for the prediction of corrosion rate and yield strength of magnesium-rare earth alloys. *Corrosion Science* 53(1):168-176.

Bishop, C. M., and N. M. Nasrabadi. 2006. *Pattern recognition and machine learning*, vol. 4. New York: Springer.

Blackwood, D. J. 2020. An electrochemist perspective of microbiologically influenced corrosion. *Corrosion and Materials Degradation* 1(1):59-76.

Borch, T., R. Kretzschmar, A. Kappler, P. V. Cappellen, M. Ginder-Vogel, A. Voegelin, and K. Campbell. 2010. Biogeochemical redox processes and their impact on contaminant dynamics. *Environmental Science & Technology* 44(1):15-23.

Boyd, B. R., J. Gattis II, W. A. Myers, and R. Selvam. 1999. *Guidelines for selections of pipe culverts*. Transportation Research Committee, Arkansas Highway and Transportation Department. http://www.ahtd.state.ar.us/TRC/TRC%20Reports/TRC9601_Guidelines_for_Selections_of_Pipe_Culverts.pdf.

Brady, K., and W. McMahon. 1994. *The durability of corrugated steel buried structures*. ARRB National Transport Research Centre.

Brenna, A., F. Bolzoni, S. Beretta, and M. Ormellese. 2017. *Can an intermittent cathodic protection system prevent corrosion of buried pipeline?* Paper presented at CORROSION 2017, New Orleans, Louisiana. Paper No. NACE-2017-9353.

Briaud, J. L., Z. Medina-Cetina, S. Hurlebaus, M. Everett, S. Tucker, N. Yousefpour. and R. Arjwech. 2012. *Unknown foundation determination for scour*. No. FHWA/TX-12/0-6604-1. College Station, TX: Texas Transportation Institute.

Broomfield, J. 2003. *Corrosion of steel in concrete: Understanding, investigation and repair*. Boca Raton, FL: CRC Press.

Caleyo, F., L. Alfonso, J. Espina-Hernández, and J. Hallen. 2007. Criteria for performance assessment and calibration of in-line inspections of oil and gas pipelines. *Measurement Science and Technology* 18(7):1787-1799.

Caleyo, F., J. C. Velázquez, A. Valor, and J. M. Hallen. 2009. Markov chain modelling of pitting corrosion in underground pipelines. *Corrosion Science* 51(9):2197-2207.

California Public Utilities Commission. 2019. Root cause analysis for Aliso Canyon finalized: California to continue strengthening safeguards for natural gas storage facilities. Press Release.

Caseres, L., and A. A. Sagues. 2005. *Corrosion of aluminized steel in scale forming waters*. Paper presented at CORROSION 2005, April 3–7. Paper No. NACE-05348X.

CEN (European Committee for Standardization) 12501-2. 2003. *Protection of metallic materials against corrosion. Corrosion likelihood in soil. Part 2: Low alloyed and unalloyed ferrous materials*. Brussels, Belgium: European Committee for Standardization.

CEN 14490. 2010. *Execution of special geotechnical works-soil nailing*. Brussels, Belgium: European Committee for Standardization.

Cerlanek, W., and R. Powers. 1993. *Drainage culvert service life performance and estimation*. Tallahassee, FL: Florida Department of Transportation.

Chen, L., B. Wei, and X. Xu. 2021. Effect of sulfate-reducing bacteria (SRB) on the corrosion of buried pipe steel in acidic soil solution. *Coatings* 11(6):625.

Cheney, R. 1988. *Permanent ground anchors: DP-68*. FHWA-DP-90-068-003. Washington, DC: Federal Highway Administration.

City of East Providence, Rhode Island. 2021. Henderson Bridge September 24, 2021, Advisory. https://eastprovidenceri.gov/news-announcements/emergency-management/henderson-bridge-september-24-2021-advisory (accessed December 1, 2021).

Clary, J., J. Jones, M. Leisenring, P. Hobson, and E. Strecker. 2020. *The International Stormwater BMP Database: 2020 summary statistics*. Alexandria, VA: Water Research Foundation.

Clouterre. 1993. *Recommendations Clouterre 1991. Soil nailing recommendations-1991 for designing, calculating, constructing and inspecting earth support systems using soil nailing*. Paris, France: Presses de l'Ecole Nationale des Ponts et Chaussees.

Coburn, S. K. 1978. Corrosion in fresh water. In *Metals handbook*. Materials Park, OH: American Society for Metals. Pp. 733-736.

Cong, H., and J. R. Scully. 2010. Use of coupled multielectrode arrays to elucidate the pH dependence of copper pitting in potable water. *Journal of the Electrochemical Society* 157(1):C36.

Corrdesa. 2022. *What is corrosion djinn?* https://www.corrdesa.com/corrosion-djinn (accessed May 20, 2022).

Dann, M. R., and M. Birkland. 2019. Structural deterioration modeling using variational inference. *Journal of Computing in Civil Engineering* 33(1):04018057.

Dann, M. R., M. A. Maes, and M. M. Salama. 2015. *Pipeline corrosion growth modeling for in-line inspection data using a population-based approach*. Paper presented at 34th International Conference on Ocean, Offshore and Arctic Engineering. American Society of Mechanical Engineers, St. John's, Newfoundland, May 31-June 5, 2015.

Darbin, M., J. Jaillaux, and J. Montuelle. 1986. La perennite des ovages en terre armee [The perenniality of the structures in armed earth—results of a long-term experiment on galvanized steel]. *Bulletin de Laison Laboratory Central des Ponts et Chaussees* 141:21-35.

Darbin, M., J. Jailloux, and J. Montuelle. 1988. Durability of reinforced earth structures: The results of a long-term study conducted on galvanized steel. *Proceedings of the Institution of Civil Engineers* 84(5):1029-1057.

de Arriba-Rodriguez, L., J. Villanueva-Balsera, F. Ortega-Fernandez, and F. Rodriguez-Perez. 2018. Methods to evaluate corrosion in buried steel structures: A review. *Metals* 8(5):334.

Decker, J. B., K. M. Rollins, and J. C. Ellsworth. 2008. Corrosion rate evaluation and prediction for piles based on long-term field performance. *Journal of Geotechnical and Geoenvironmental Engineering* 134(3):341-351.

DIN (Deutsches Institut für Normung) 50929-3. 1985. Probability of corrosion of metallic materials when subject to corrosion from the outside. In *Buried and underwater pipelines and structural components*. Berlin, Germany: German Institute for Standardization.

DMRB (Design Manual for Roads and Bridges) BD 42/00. 2000. *Highway structures: Design, part 2*, vol. 2. Guildford, UK: UK Highways Agency.

DNV (Det Norske Veritas). 2010. Cathodic protection design. Recommended Practice DNV-RP-B401. Høvik, Norway: DNV.

Drumm, E. C., N. R. Rainwater, W. C. Wright, and R. E. Yoder. 2006. Evaluation of instrumentation for monitoring seasonal variations in pavement subgrade water content. In *TRB 85th Annual Meeting compendium of papers* [CD-ROM]. Washington, DC: Transportation Research Board.

DVGW (Deutsche Verein des Gas- und Wasserfaches) GW 9. 2011. *Evaluation of soils in view of their corrosion behavior towards buried pipelines and vessels of non-alloyed iron materials*. Bonn, Germany: German Technical and Scientific Association for Gas and Water.

Eckert, R. 2003. *Field guide for investigating internal corrosion of pipelines*. Pittsburgh, PA: National Association of Corrosion Engineers Press.

Eguchi, R. T., H. A. Seligson, and D. G. Honegger. 1995. *Pipeline replacement feasibility study: A methodology for minimizing seismic and corrosion risks to underground natural gas pipelines*. MCEER Technical Report. University at Buffalo, State University of New York.

Eid, M. M., K. E. Duncan, and R. S. Tanner. 2018. A semi-continuous system for monitoring microbially influenced corrosion. *Journal of Microbiological Methods* 150:55-60.

Elias, V. 1990. Corrosion of metallic reinforcement. In *Durability/corrosion of soil reinforced structures*. FHWA-RD-89-186. McLean, VA: Federal Highway Administration. P. 36.

Elsyca. 2022. Corrosion modeling. https://www.elsyca.com/innovate/corrosion-modeling (accessed May 20, 2022).

Engelhardt, G., D. Macdonald, and M. Urquidi-Macdonald. 1999. Development of fast algorithms for estimating stress corrosion crack growth rate. *Corrosion Science* 41(12):2267-2302.

Enning, D., and J. Garrelfs. 2014. Corrosion of iron by sulfate-reducing bacteria: New views of an old problem. *Applied Environmental Microbiology* 80(4):1226-1236.

EPA (U.S. Environmental Protection Agency). 2015. Galena trail derailment. https://19january2017snapshot.epa.gov/il/galena-train-derailment_.html (accessed December 1, 2021).

Evans, U., R. Mears, and P. Queneau. 1933. Corrosion velocity and corrosion probability. *Engineering* 136:689.

FDOT (Florida Department of Transportation). 2018. Structures design guidelines. In *Structures manual*, vol. 1. Tallahassee, FL: FDOT.

Fédération Internationale de la Précontrainte. 1986. Corrosion and corrosion protection of prestressed ground anchorages, FIP State of the Art Report. London, UK: Thomas Telford.

Ferris, F. G., L. G. Stehmeier, A. Kantzas, and F. M. Mourits. 1996. Bacteriogenic mineral plugging. *Journal of Canadian Petroleum Technology* 35(8).

Finneran, S., B. Krebs, and L. Bensman. 2015. *Criteria for pipelines co-existing with electric power lines*. Final Report No. 2015-04. Washington, DC: INGAA Foundation. P. 66.

Fishman, K. L. 2005. *Phase II: Condition assessment and evaluation of rock reinforcement along I-93 Barron Mountain rock cut, Woodstock, NH: Validation of NDT results for condition assessment of rock reinforcements*. Final Report. FHWA-NH-RD-14282C. Springfield, VA: National Technical Information Service.

Fishman, K. L., and R. Richards. 1997. *Seismic analysis and design of bridge abutments considering sliding and rotation*. NCEER-97-0009. Springfield, VA: National Technical Information Service.

Fishman, K. L., and J. L. Withiam. 2011. *NCHRP Report 675: LRFD metal loss and service-life strength reduction factors for metal-reinforced systems*. Washington, DC: Transportation Research Board.

Fishman, K. L., S. Nazarian, S. Walker, and A. Bronson. 2021. *NCHRP Research Report 958: Electrochemical test methods to evaluate the corrosion potential of earthen materials*. Washington, DC: Transportation Research Board.

Fletcher, F. B. 2005. Corrosion of weathering steel. In *ASM handbook, volume 13B: Corrosion: Materials*, edited by D. C. Cramer and B. S. Covino, Jr. Materials Park, OH: ASM International.

Flounders, E. J., and D. Lindemuth. 2015. *Development and testing of a linear polarization resistance corrosion rate probe for ductile iron pipe* (Web Report 4361). Denver, CO: Water Research Foundation. https://www.waterrf.org/research/projects/development-and-testing-linear-polarization-resistance-corrosion-rate-probe (accessed January 25, 2023).

Flounders, E. C., and S. A. Memon. 2020. *TCRP Research Report 212: Stray current control of direct current-powered rail transit systems: A guidebook.* Washington, DC: Transportation Research Board.

Fontes, E., and B. Nistad. 2019. *Modeling corrosion and corrosion protection.* White Paper. COMSOL Inc.

Frankel, G. S., T. Li, and J. R. Scully. 2017. Localized corrosion: Passive film breakdown vs pit growth stability. *Journal of the Electrochemical Society* 164(4):C180-C181.

Gadd, G. M. 2010. Metals, minerals and microbes: Geomicrobiology and bioremediation. *Microbiology* 156(3):609-643.

Genc, D., J. C. Ashlock, B. Cetin, and P. Kremer. 2019. Development and pilot installation of a scalable environmental sensor monitoring system for freeze–thaw monitoring under granular-surfaced roadways. *Transportation Research Record* 2673(12):880-890.

Gladstone, R. A., P. L. Anderson, K. L. Fishman, and J. L. Withiam. 2006. Durability of galvanized soil reinforcement: More than 30 years of experience with mechanically stabilized earth. *Transportation Research Record* 1975(1):49-59.

Gong, C., and W. Zhou. 2017a. First-order reliability method-based system reliability analyses of corroding pipelines considering multiple defects and failure modes. *Structure and Infrastructure Engineering* 13(11):1451-1461.

Gong, C., and W. Zhou. 2017b. Improvement of equivalent component approach for reliability analyses of series systems. *Structural Safety* 68:65-72.

Gong, C., and W. Zhou. 2018a. Importance sampling-based system reliability analysis of corroding pipelines considering multiple failure modes. *Reliability Engineering & System Safety* 169:199-208.

Gong, C., and W. Zhou. 2018b. Multi-objective maintenance strategy for in-service corroding pipelines using genetic algorithms. *Structure and Infrastructure Engineering* 14(11):1561-1571.

Gong, J., P. W. Jayawickrama, and Y. Tinkey. 2006. Nondestructive evaluation of installed soil nails. *Transportation Research Record* 1(1):104-113.

Gorbushina, A. A. 2007. Life on the rocks. *Environmental Microbiology* 9(7):1613-1631.

Greene, N. D., and M. G. Fontana. 1959. A critical analysis of pitting corrosion. *Corrosion* 15(1):41-47.

Grundl, T. J., S. Haderlein, J. T. Nurmi, and P. G. Tratnyek. 2011. Introduction to aquatic redox chemistry. *Aquatic redox chemistry*, edited by P. G. Tratnyek, T. J. Grundl, and S. B. Haderlein. Washington, DC: American Chemical Society.

Gu, L., J. H. Fenton, S. J. Welsh, A. G. Mouradian, and S. E. McInnes. 2015. Corrosion study of existing steel H-pile installed through cinder-ash fill. In *IFCEE 2015: Proceedings of the International Foundations Congress and Equipment Expo.* Alexandria, VA: American Society of Civil Engineers. Pp. 961-972.

Gu, T. 2014. Theoretical modeling of the possibility of acid producing bacteria causing fast pitting biocorrosion. *Journal of Microbial & Biochemical Technology* 6(2):68-74.

Gu, T., and B. Galicia. 2012. *Can acid producing bacteria be responsible for very fast MIC pitting?* Paper presented at CORROSION/2012, Salt Lake City, Utah, March 11-15.

Guan, F., X. Zhai, J. Duan, M. Zhang, and B. Hou. 2016. Influence of sulfate-reducing bacteria on the corrosion behavior of high strength steel EQ70 under cathodic polarization. *PloS One* 11(9):e0162315.

Gumbel, E. J. 1954. *Statistical theory of extreme values and some practical applications.* National Bureau of Standards Applied Mathematics Series 33. Washington, DC: U.S. Government Printing Office.

Gumbel, E. J. 2004. *Statistics of extremes.* Reprint of the 1958 edition: Mineola, NY: Dover Publications. P. 375.

Haïun, G., J. Jailloux, and F. Renaudin. 2007. Bilan des investigations effectuées sur des ouvrages en terre armée [Assessment of the investigations carried out on reinforced earth structures]. *SETRA-Ouvrages d'Art* 1(55):31-35.

Heckerman, D. 2008. A tutorial on learning with Bayesian networks. *Innovations in Bayesian networks*, vol. 156, edited by D. E. Holmes and L. C. Jain. Berlin, Heidelberg: Springer. Pp. 33-82.

Helsel, J. L. 2018. *Expected service life and cost considerations for maintenance and new construction protective coating work.* Paper presented at CORROSION 2018, Phoenix, Arizona. Paper No. NACE-2018-10673.

Hinshaw, R. F., and J. Northrup. 1991. Predicting subgrade moisture under aggregate surfacing. *Transportation Research Record* 1291:193-203.

Hiromoto, S. 2010. Corrosion of metallic biomaterials. In *Metals for biomedical devices.* Sawston, UK: Woodhead Publishing. Pp. 99-121.

Hong, H. 1999. Inspection and maintenance planning of pipeline under external corrosion considering generation of new defects. *Structural Safety* 21(3):203-222.

Hördt, A., R. Blaschek, A. Kemna, and N. Zisser. 2007. Hydraulic conductivity estimation from induced polarisation data at the field scale—the Krauthausen case history. *Journal of Applied Geophysics* 62(1):33-46.

Howard, G. T. 2011. Microbial biodegradation of polyurethane. In *Recent developments in polymer recycling*, edited by A. Fainleib and O. Grigoryeva. Kerala, India: Transworld Research Network. Pp. 215-238.

Hubbard, S. S., J. Zhang, P. J. Monteiro, J. E. Peterson, and Y. Rubin. 2003. Experimental detection of reinforcing bar corrosion using nondestructive geophysical techniques. *ACI Materials Journal* 100(6):501-510.

IEEE Std 81. 2012. *IEEE guide for measuring earth resistivity, ground impedance, and earth surface potentials of a grounding system*. (Revision of IEEE Std 81-1983.) Piscataway, NJ: IEEE.

Islam, M., A. R. Al-Shamari, S. Al-Sulaiman, S. Prakas, A. Jaragh, and S. Abraham. 2016. *Characteristic corrosion morphological features associated with different strains of bacteria species in oilfield waters*. Paper presented at CORROSION 2016, Vancouver, British Columbia.

Jackura, J., G. Garofalo, and D. Beddard. 1987. *Investigation of corrosion at 14 mechanically stabilized embankment sites*. Office of Transportation Laboratory, California Department of Transportation.

Jansen, S., M. van Burgel, J. Gerritse, and M. Büchler. 2017. *Cathodic protection and MIC—effects of local electrochemistry*. Paper presented at CORROSION 2017, New Orleans, Louisiana. Paper No. NACE-2017-9452.

Ji, J., D. Robert, C. Zhang, D. Zhang, and J. Kodikara. 2017. Probabilistic physical modelling of corroded cast iron pipes for lifetime prediction. *Structural Safety* 64:62-75.

Johnell, K., and I. Klarin. 2007. The relationship between number of drugs and potential drug–drug interactions in the elderly. *Drug Safety* 30(10):911-918.

Kamrunnahar, M., and M. Urquidi-Macdonald. 2006. Data mining of experimental corrosion data using neural network. *ECS Transactions* 1(4):71-79.

Kamrunnahar, M., and M. Urquidi-Macdonald. 2010. Prediction of corrosion behavior using neural network as a data mining tool. *Corrosion Science* 52(3):669-677.

Kamrunnahar, M., and M. Urquidi-Macdonald. 2011. Prediction of corrosion behaviour of alloy 22 using neural network as a data mining tool. *Corrosion Science* 53(3):961-967.

Kappler, A., C. Bryce, M. Mansor, U. Lueder, J. M. Byrne, and E. D. Swanner. 2021. An evolving view on biogeochemical cycling of iron. *Nature Reviews Microbiology* 19(6):360-374.

Karim, M. Z., S. E. Tucker-Kulesza, and M. Bernhardt-Barry. 2019. Electrical resistivity as a binary classifier for bridge scour evaluation. *Transportation Geotechnics* 19:146-157.

Ke, H., and C. D. Taylor. 2019. Density functional theory: An essential partner in the integrated computational materials engineering approach to corrosion. *Corrosion* 75(7):708-726.

Keller Management Services, LLC. 2022. Techniques. https://www.keller-na.com/expertise/techniques (accessed August 24, 2022).

Kemna, A. 2000. *Tomographic inversion of complex resistivity: Theory and application*. Osnabrück, Germany: Der Andere Verlag.

Kendorski, F. S. 2000. Site characterization for planning underground stone mines. In *Proceedings of the 19th International Conference on Ground Control in Mining, Morgantown, WV*, edited by S. S. Peng and C. Mark. Morgantown, WV: West Virginia University. Pp. 137-148.

Kennelley, K., L. Bone, and M. Orazem. 1993. Current and potential distribution on a coated pipeline with holidays, part I—Model and experimental verification. *Corrosion* 49(3):199-210.

Kessouri, P., A. Furman, J. A. Huisman, T. Martin, A. Mellage, D. Ntarlagiannis, M. Bücker, S. Ehosioke, P. Fernandez, A. Flores-Orozco, and A. Kemna. 2019. Induced polarization applied to biogeophysics: Recent advances and future prospects. *Near Surface Geophysics* 17(6):595-621.

Khanna, A. S. 2018. *Oil & gas corrosion protection—underground pipeline protection strategies*. World of Chemicals. https://www.worldofchemicals.com/media/oil-gas-corrosion-protection-underground-pipeline-protection-strategies/967.html (accessed July 14, 2022).

Kiani Khouzani, M., A. Bahrami, A. Hosseini-Abari, M. Khandouzi, and P. Taheri. 2019. Microbiologically influenced corrosion of a pipeline in a petrochemical plant. *Metals* 9(4):459.

Kieft, T. L. 2000. Size matters: Dwarf cells in soil and subsurface terrestrial environments. In *Nonculturable microorganisms in the environment*. Boston, MA: Springer. Pp. 19-46.

Kim, C., L. Chen, H. Wang, and H. Castaneda. 2021. Global and local parameters for characterizing and modeling external corrosion in underground coated steel pipelines: A review of critical factors. *Journal of Pipeline Science and Engineering* 1(1):17-35.

Kim, T.-G. 2002. Failure of piping by hydrogen attack. *Engineering Failure Analysis* 9(5):571-578.

King, R. 1977. *A review of soil corrosiveness with particular reference to reinforced earth: Transport and Road Research Lab Supplementary Report 316*. Wokingham, Berkshire: Transport Research Library.

Klindworth, A., E. Pruesse, T. Schweer, J. Peplies, C. Quast, M. Horn, and F. O. Glöckner. 2013. Evaluation of general 16s ribosomal RNA gene PCR primers for classical and next-generation sequencing-based diversity studies. *Nucleic Acids Research* 41(1):e1.

Knight, R. J., and A. L. Endres. 2005. An introduction to rock physics principles for near-surface geophysics. In *Near-surface geophysics. Houston, TX:* Society of Exploration Geophysicists. Pp. 31-70.

Koch, G. 2017. Cost of corrosion. In *Trends in oil and gas corrosion research and technologies*, edited by A. M. El-Sherik. Boston, MA: Woodhead Publishing. Pp. 3-30.

Korinko, P., T. Adams, and G. Creech. 2005. *Hydrogen permeation resistant coatings*. Paper presented at Materials Science & Technology, Pittsburgh, PA.

Kramer, S. L. 1993. *Evaluation of tieback performance: Final technical report*. Seattle, WA: Washington State Department of Transportation.

Kuhn, R. 1928. Galvanic current on cast iron pipes. Their causes and effects. Methods of measuring and means of prevention. In *Proceedings, Soil Corrosion Conference*. Washington, DC: Bureau of Standards.

LADWP (Los Angeles Department of Water and Power). 2016. *Water infrastructure plan*. https://www.ladwp.com/ladwp/faces/ladwp/aboutus/a-water/a-w-project/a-w-p-infrastructureimprovement (accessed January 24, 2023).

Laycock, P., R. Cottis, and P. Scarf. 1990. Extrapolation of extreme pit depths in space and time. *Journal of the Electrochemical Society* 137(1):64.

Lazarte, C. A., V. Elias, R. D. Espinoza, and P. J. Sabatini. 2003. Soil nail walls. *Geotechnical Engineering Circular No. 7*. Report Number FHWA0-IF-03-017.

Lee, S. I., D. G. Zollinger, R. L. Lytton, and N. C. Jackson. 2009. Automation of pavement sublayer moisture content determination using long-term pavement performance time domain reflectometry data and micromechanics. *Transportation Research Record* 2116(1):16-25.

Leeds, S., and J. Leeds. 2015. Cathodic protection. In *Oil and gas pipelines*, edited by R. W. Revie. Hoboken, NJ: John Wiley & Sons. Pp. 457-484.

Li, S., Y. Kim, K. Jeon, Y. Kho, and T. Kang. 2001. Microbiologically influenced corrosion of carbon steel exposed to anaerobic soil. *Corrosion* 57(9):815-828.

Li, S. Y., Y. G. Kim, K. S. Jeon, and Y. T. Kho. 2000. Microbiologically influenced corrosion of underground pipelines under the disbonded coatings. *Metals and Materials* 6(3):281-286.

Li, T., G. S. Frankel, and J. R. Scully. 2018a. Localized corrosion: Passive film breakdown vs. pit growth stability: Part III. A unifying set of principal parameters and criteria for pit stabilization and salt film formation. *Journal of the Electrochemical Society* 167(10):C762-C770.

Li, T., J. R. Scully, and G. S. Frankel. 2018b. Localized corrosion: Passive film breakdown vs pit growth stability: Part II. A model for critical pitting temperature. *Journal of the Electrochemical Society* 165(9):C484.

Li, T., J. R. Scully, and G. S. Frankel. 2019a. Localized corrosion: Passive film breakdown vs. Pit growth stability: Part IV. The role of salt film in pit growth: A mathematical framework. *Journal of the Electrochemical Society* 166(6):C115-C124.

Li, T., J. R. Scully, and G. S. Frankel. 2019b. Localized corrosion: Passive film breakdown vs pit growth stability: Part V. Validation of a new framework for pit growth stability using one-dimensional artificial pit electrodes. *Journal of the Electrochemical Society* 166(11):C3341-C3354.

Li, T., J. Wu, and G. S. Frankel. 2021. Localized corrosion: Passive film breakdown vs. pit growth stability, Part VI: Pit dissolution kinetics of different alloys and a model for pitting and repassivation potentials. *Corrosion Science* 182:109277.

Li, X., and H. Castaneda. 2015. Coating studies of buried pipe in soil by novel approach of electrochemical impedance spectroscopy at wide frequency domain. *Corrosion Engineering, Science and Technology* 50(3):218-225.

Li, X., and H. Castaneda. 2017. Damage evolution of coated steel pipe under cathodic-protection in soil. *Anti-Corrosion Methods and Materials* 64(1):118-126.

Li, X., O. Rosas, and H. Castaneda. 2016. Deterministic modeling of API5l X52 steel in a coal-tar-coating/cathodic-protection system in soil. *International Journal of Pressure Vessels and Piping* 146:161-170.

Li, Y., D. Xu, C. Chen, X. Li, R. Jia, D. Zhang, W. Sand, F. Wang, and T. Gu. 2018. Anaerobic microbiologically influenced corrosion mechanisms interpreted using bioenergetics and bioelectrochemistry: A review. *Journal of Materials Science & Technology* 34(10):1713-1718.

Liang, R. Y., K. Al-Akhras, and S. Rabab'ah. 2006. Field monitoring of moisture variations under flexible pavement. *Transportation Research Record* 1967(1):160-172.

Little, B. J., and J. S. Lee. 2007. *Microbiologically influenced corrosion*, vol. 3. New York: John Wiley & Sons.

Little, B. J., and J. S. Lee. 2018. Microbiologically influenced corrosion. In *Peabody's control of pipeline corrosion*, edited by R. Bianchetti. Houston, TX: NACE International.

Little, B. J., and P. A. Wagner. 2002. Application of electrochemical techniques to the study of microbiologically influenced corrosion. In *Modern aspects of electrochemistry*, edited by J. O. Bockris, B. E. Conway, and R. E. White. Boston, MA: Springer. Pp. 205-246.

Little, B. J., D. Blackwood, J. Hinks, F. Lauro, E. Marsili, A. Okamoto, S. Rice, S. Wade, and H.-C. Flemming. 2020. Microbially influenced corrosion—any progress? *Corrosion Science* 170:108641.

Littlejohn, G. 1992. Advancing anchorage technology. *International Journal of Rock Mechanics and Mining Sciences & Geomechanics Abstracts* 62(7):61-64.

Loehr, J. E., A. Lutenegger, B. L. Rosenblad, A. Boeckmann, and P. Brinckerhoff. 2016. *Geotechnical site characterization*. Geotechnical Engineering Circular No. 5. https://rosap.ntl.bts.gov/view/dot/40558 (accessed January 24, 2023).

Loewer, M., T. Günther, J. Igel, S. Kruschwitz, T. Martin, and N. Wagner. 2017. Ultra-broad-band electrical spectroscopy of soils and sediments—a combined permittivity and conductivity model. *Geophysical Journal International* 210(3):1360-1373.

Logan, K. 1945. *Underground corrosion*. National Bureau of Standards Circular 450. Washington, DC: U.S. Department of Commerce.

Loi, J., D. Fredlund, J. Gan, and R. Widger. 1992. Monitoring soil suction in an indoor test track facility. *Transportation Research Record* 1362:101-110.

Lokse, A. 1992. *Durability and long-term performance of rock bolts*. Trondheim, Norway: Norwegian Institute of Technology, University of Trondheim.

Long, R. P. 1992. Corrosion of steel piles in some waste fills. *Transportation Research Record* 1345:53-59.

Long, R. P., J. Badinter, and P. R. Kambala. 1995. *Investigation of steel pile foundations in corrosively active areas: Final report*. Mansfield, CT: University of Connecticut.

Maes, M. A., M. H. Faber, and M. R. Dann. 2009. *Hierarchical modeling of pipeline defect growth subject to ILI uncertainty*. Paper presented at 28th International Conference on Offshore Mechanics and Arctic Engineering. American Society of Mechanical Engineers, Honolulu, Hawaii, May 31-June 5, 2009.

Makama, Z., S. Celikkol, A. Ogawa, C. Gaylarde, and I. Beech. 2018. The issue with using DNA profiling as a sole method for investigating the role of marine biofilms in corrosion of metallic materials. *International Biodeterioration & Biodegradation* 135:33-38.

Mand, J., H. S. Park, T. R. Jack, and G. Voordouw. 2014. The role of acetogens in microbially influenced corrosion of steel. *Frontiers in Microbiology* 5:268.

Martín, Ó., P. De Tiedra, and M. López. 2010. Artificial neural networks for pitting potential prediction of resistance spot welding joints of AISI 304 austenitic stainless steel. *Corrosion Science* 52(7):2397-2402.

McCarthy, S., J. Neatrour, and J. English. 1988. Attitudes and practices: Direct current transit systems and stray current corrosion. *Transportation Research Record* 1162:34-42.

McCartney, J. S., and A. Khosravi. 2013. Field-monitoring system for suction and temperature profiles under pavements. *Journal of Performance of Constructed Facilities* 27(6):818-825.

Mears, R., and U. Evans. 1935. The "probability" of corrosion. *Transactions of the Faraday Society* 31:527-542.

Mehta, P., and P. Monteiro. 2014. *Concrete: Microstructure, properties, and materials*. New York: McGraw-Hill Education.

Melchers, R. 2003. Modeling of marine immersion corrosion for mild and low-alloy steels, Part 1: Phenomenological model. *Corrosion* 59(4):319-334.

Melchers, R. 2004. Pitting corrosion of mild steel in marine immersion environment, Part 1: Maximum pit depth. *Corrosion* 60(9):824-836.

Melchers, R. E. 2005a. Representation of uncertainty in maximum depth of marine corrosion pits. *Structural Safety* 27(4):322-334.

Melchers, R. E. 2005b. Statistical characterization of pitting corrosion, Part 1: Data analysis. *Corrosion* 61(7):655-664.

Melchers, R. E. 2005c. Statistical characterization of pitting corrosion, Part 2: Probabilistic modeling for maximum pit depth. *Corrosion* 61(8):766-777.

Melchers, R. E. 2008. Extreme value statistics and long-term marine pitting corrosion of steel. *Probabilistic Engineering Mechanics* 23(4):482-488.

Melchers, R. E. 2010. Estimating uncertainty in maximum pit depth from limited observational data. *Corrosion Engineering, Science and Technology* 45(3):240-248.

Melchers, R. E. 2015. Progression of pitting corrosion and structural reliability of welded steel pipelines. In *Oil and gas pipelines: Integrity and safety handbook*, edited by R. W. Revie. New York: John Wiley & Sons. Pp. 327-342.

Melchers, R. E. 2020. Models for prediction of long-term corrosion of cast iron water mains. *Corrosion* 76(5):441-450.

Melchers, R. E., and A. T. Beck, eds. 2018. *Structural reliability analysis and prediction*. New York: John Wiley & Sons.

Melchers, R. E., and R. Petersen. 2018. A reinterpretation of the Romanoff NBS data for corrosion of steels in soils. *Corrosion Engineering, Science and Technology* 53(2):131-140.

MIL-STD-889D. 2016. *Galvanic compatibility of electrically conductive materials*. Washington, DC: U.S. Department of Defense.

Mitchell, J. K., and K. Soga. 2005. *Fundamentals of soil behavior*, vol. 3. New York: John Wiley & Sons.

Miyanaga, K., R. Terashi, H. Kawai, H. Unno, and Y. Tanji. 2007. Biocidal effect of cathodic protection on bacterial viability in biofilm attached to carbon steel. *Biotechnology & Bioengineering* 97(4):850-857.

Mohandas, K., and D. Fray. 2011. Novel electrochemical measurements on direct electro-deoxidation of solid TiO_2 and ZrO_2 in molten calcium chloride medium. *Journal of Applied Electrochemistry* 41(3):321-336.

Moody, W. 1993. *Corrosion in the soil environment: New York's experience*. Paper presented at 13th Central Pennsylvania Geotechnical Seminar: Progress in Geotechnical Engineering Practice, Hershey, PA. American Society of Civil Engineers.

Mujah, D., M. A. Shahin, and L. Cheng. 2017. State-of-the-art review of biocementation by microbially induced calcite precipitation (MICP) for soil stabilization. *Geomicrobiology Journal* 34(6):524-537.

NACE (National Association of Corrosion Engineers International). 2001. State-of-the-Art Survey on Corrosion of Steel Piling in Soils. NACE 05101. https://store.ampp.org/05101-state-of-the-art-survey-on-corrosion-of-steel-piling-in-soils (accessed January 24, 2023).

NACE Committee TEG 187X. 2019. *Microbiologically influenced corrosion*. Paper presented at CORROSION 2019, Nashville, TN.

NACE SP0169. 2013. *Control of external corrosion on underground or submerged metallic piping systems*. Houston, TX: NACE International.

NACE TM0106. 2016. *Detection, testing, and evaluation of microbiologically influenced corrosion (MIC) on external surfaces of buried pipelines*. Houston, TX: NACE International.

NASEM (National Academies of Sciences, Engineering, and Medicine). 2011. *Validation of LRFD metal loss and service-life strength reduction factors for metal-reinforced systems*. NCHRP Research Results Digest 364. Washington, DC: The National Academies Press.

NASEM. 2020. *Life-cycle decisions for biomedical data: The challenge of forecasting costs*. Washington, DC: The National Academies Press.

NCHRP (National Cooperative Highway Research Program). 2017 (unpublished). *Improved test methods and practices for characterizing steel corrosion potential of earthen materials, phase I*.

Nealson, K. H., and D. A. Stahl. 1997. Microorganisms and biogeochemical cycles: What can we learn from layered microbial communities? *Reviews in Mineralogy and Geochemistry* 35(1):5-34.

Nemati, M., and G. Voordouw. 2003. Modification of porous media permeability, using calcium carbonate produced enzymatically in situ. *Enzyme and Microbial Technology* 33(5):635-642.

New Zealand Geotechnical Society Inc. 2017. Electronic transfer of geotechnical and geoenvironmental data. Vol. AGS4 NZ v1.0.1. https://fl-nzgs-media.s3.amazonaws.com/uploads/2016/06/AGS-NZ-V4.0.1-October-2017.pdf (accessed February 1, 2023).

NHERI (Natural Hazards Engineering Research Infrastructure). 2022. *About NHERI DesignSafe*. https://www.designsafe-ci.org/about/designsafe/ (accessed May 20, 2022).

Nicks, J. E., M. T. Adams, and M. Runion. 2015. Electronic thickness gauge measurements on a 36-year-old steel bar mat reinforced MSE wall. In *IFCEE 2015*, edited by M. Iskander, M. T. Suleiman, J. B. Anderson, and D. F. Laefer. Reston, VA: American Society of Civil Engineers. Pp. 636-647.

Nicks, J. E., M. T. Adams, T. Stabile, and J. Ocel. 2017. Case study: Condition assessment of a 36-year-old mechanically stabilized earth wall in Virginia. *Journal of Geotechnical and Geoenvironmental Engineering* 143(5):05016003.

NRC (National Research Council). 2009. *Review of the Bureau of Reclamation's corrosion prevention standards for ductile iron pipe*. Washington, DC: The National Academies Press.

NSF (National Science Foundation). 2020. *Engineering research centers*. https://www.nsf.gov/eng/eec/erc.jsp (accessed March 29, 2022).

NSF. 2021a. *Accelerating impact through partnerships: Industry-university cooperative research centers (IUCRC)*. https://iucrc.nsf.gov (accessed Oct 26, 2021).

NSF. 2021b. *Science and technology centers (STCS): Integrative partnerships*. https://www.nsf.gov/od/oia/programs/stc (accessed October 26, 2021).

Ohsaki, Y. 1982. Corrosion of steel piles driven in soil deposits. In *Soil and foundations*. Tokyo, Japan: Japanese Geotechnical Society. Pp. 57-76.

Oldfield, J. W., and W. H. Sutton. 1978. Crevice corrosion of stainless steels: I. A mathematical model. *British Corrosion Journal* 13(1):13-22.

OLI Systems. 2022. *Enhance process research, design and operations performance with rigorous chemistry analysis.* https://www.olisystems.com/software/oli-studio (accessed May 20, 2022).

Oparaodu, K. O., and G. C. Okpokwasili. 2014. Effects of tetrakis (hydroxymethyl) phosphorium sulphate biocides on metal loss in mild steel coupon buried in a water-logged soil. *Journal of Applied and Environmental Microbiology* 2:253-256.

O'Rourke, M. J., and X. Liu. 2012. *Seismic design of buried and off-shore pipelines.* MCEER-11-MN04. MCEER, University at Buffalo, State University of New York.

Ortiz, M., P. M. Leung, G. Shelley, T. Jirapanjawat, P. A. Nauer, M. W. Van Goethem, S. K. Bay, Z. F. Islam, K. Jordaan, and S. Vikram. 2021. Multiple energy sources and metabolic strategies sustain microbial diversity in Antarctic desert soils. *Proceedings of the National Academy of Sciences of the United States of America* 118(45):e2025322118.

Ossai, C. I. 2020. Corrosion defect modelling of aged pipelines with a feed-forward multi-layer neural network for leak and burst failure estimation. *Engineering Failure Analysis* 110:104397.

Palani, S., T. Hack, J. Deconinck, and H. Lohner. 2014. Validation of predictive model for galvanic corrosion under thin electrolyte layers: An application to aluminum 2024-CFRP material combination. *Corrosion Science* 78:89-100.

Paszke, A., S. Gross, F. Massa, A. Lerer, J. Bradbury, G. Chanan, T. Killeen, Z. Lin, N. Gimelshein, L. Antiga, A. Desmaison, A. Köpf, E. Yang, Z. DeVito, M. Raison, A. Tejani, S. Chilamkurthy, B. Steiner, L. Fang, J. Bai, and S. Chintala. 2019. Pytorch: An imperative style, high-performance deep learning library. In *Advances in neural information processing systems 32*, edited by H. M. Wallach, H. Larochelle, A. Beygelzimer, F. d'Alché-Buc, E. A. Fox, and R. Garnett. Pp. 8024-8035. https://researchr.org/publication/nips-2019 (accessed January 24, 2023).

Pedregosa, F., G. Varoquaux, A. Gramfort, V. Michel, B. Thirion, O. Grisel, M. Blondel, P. Prettenhofer, R. Weiss, and V. Dubourg. 2011. Scikit-learn: Machine learning in python. *Journal of Machine Learning Research* 12:2825-2830.

Pelecanos, L., and K. Soga. 2017. *Innovative structural health monitoring of foundation piles using distributed fibre-optic sensing.* Paper presented at 8th International Conference on Structural Engineering and Construction Management, Sydney, Australia, November 23-28, 2015.

Pope, D. H. 1990. *Microbiologically influenced corrosion (MIC): Methods of detection in the field.* GRI field guide. GRI-90/0299. Chicago, IL: Gas Research Institute.

Pope, D. H., and E. Morris III. 1995. Some experiences with microbiologically influenced corrosion of pipelines. *Materials Performance* 34(5).

Pope, D. H., T. P. Zintel, O. W. Siebert, and T. P. Kuruvilla. 1988. *Organic acid corrosion of carbon steel—a mechanism of microbiologically influenced corrosion.* Paper presented at CORROSION/88, St. Louis, MO.

Post-Tensioning Institute. 2014. *Recommendations for prestressed rock and soil anchors,* 5th ed. Farmington Hills, MI: Post-Tensioning Institute.

Pourbaix, M. 1974. *Atlas of electrochemical equilibria in aqueous solution.* Houston, TX: National Association of Corrosion Engineers International.

Poursaee, A., P. Rangaraju, and L. Ding. 2019. *Evaluation of H-pile corrosion rates for WI bridges located in aggressive subsurface environments.* Wisconsin Highway Research Program WHRP 0092-16-03. Madison, WI: Wisconsin Department of Transportation.

Power, M., K. Fishman, F. Makdisi, S. Musser, R. Richards, and T. L. Youd. 2006. *Seismic retrofitting manual for highway structures: Part 2—Retaining structures, slopes, tunnels, culverts and roadways.* MCEER-06-SP11. University at Buffalo, State University of New York.

Provan, J., and E. Rodriguez III. 1989. Part I: Development of a Markov description of pitting corrosion. *Corrosion* 45(3):178-192.

Rachman, A., T. Zhang, and R. M. C. Ratnayake. 2021. Applications of machine learning in pipeline integrity management: A state-of-the-art review. *International Journal of Pressure Vessels and Piping* 193:104471.

Rahman, N., and M. Ismail. 2013. Corrosion protection coating for buried pipelines: A short review. *International Journal of Material Science Innovations* 4(4):207-217.

Rankin, L., and H. Al Mahrous. 2005. *External corrosion probability assessment for carrier pipes inside casings.* GRI-05/0020. Des Plaines, IL: Gas Technology Institute.

Ricker, R. E. 2010. Analysis of pipeline steel corrosion data from NBS (NIST) studies conducted between 1922–1940 and relevance to pipeline management. *Journal of Research of the National Institute of Standards and Technology* 115(5):373.

Roberge, P. 2000. *Handbook of corrosion engineering.* New York: McGraw-Hill Education.

Rodriguez, E., and J. Provan. 1989. Part II: Development of a general failure control system for estimating the reliability of deteriorating structures. *Corrosion* 45(3):193-206.

Romanoff, M. 1957. Underground corrosion. National Bureau of Standards Circular 579. Washington, DC: U.S. Government Printing Office.

Romanoff, M. 1964. Exterior corrosion of cast-iron pipe. *Journal of the American Water Works Association* 56(9):1129-1143.

Rosen, E., and D. Silverman. 1992. Corrosion prediction from polarization scans using an artificial neural network integrated with an expert system. *Corrosion* 48(9):734-745.

Sagues, A. A., N. D. Poor, L. Caseres, and M. Akhoondan. 2009. *Development of a rational method for predicting corrosion rates of metals in soils and water: Final report*, October 31. Sarasota, FL: University of South Florida.

Sankey, J. E., and D. Hutchinson. 2011. *Experience with support of railways using mechanically stabilized earth in the USA*. Paper presented at GEORAIL-2011 International Symposium on Railway Geotechnical Engineering, Paris, France.

Santamarina, J. C., V. A. Rinaldi, D. Fratta, K. A. Klein, Y.-H. Wang, G. C. Cho, and G. Cascante. 2005. A survey of elastic and electromagnetic properties of near-surface soils. *Near-Surface Geophysics* 1:71-87.

Sarhan, H. A., M. W. O'Neill, and P. D. Simon. 2002. Corrosion of reinforcing steel in drilled shafts with construction flaws. *Transportation Research Record* 1786(1):96-103.

Sasaki, T., J. Park, K. Soga, T. Momoki, K. Kawaguchi, H. Muramatsu, Y. Imasato, A. Balagopal, J. Fontenot, and T. Hall. 2019. Distributed fibre optic strain sensing of an axially deformed well model in the laboratory. *Journal of Natural Gas Science and Engineering* 72:103028.

Schwerdtfeger, W. J., and M. Romanoff. 1972. NBS papers on underground corrosion of steel piling 1962-1971. NBS Monograph 127. Washington, DC: U.S. Department of Commerce.

Sempra. 2019. *SoCalGas issues statement on Blade's analysis of Aliso Canyon well failure*. PRNewswire, May 17. https://www.sempra.com/socalgas-issues-statement-blades-analysis-aliso-canyon-well-failure (accessed July 14, 2022).

Shakoori, A. R. 2017. Fluorescence in situ hybridization (fish) and its applications. In *Chromosome structure and aberrations*. New Delhi, India: Springer India. Pp. 343-367.

Shibata, T. 1991. Evaluation of corrosion failure by extreme value statistics. *ISIJ International* 31(2):115-121.

Shibata, T. 1994. Application of extreme-value statistics to corrosion. *Journal of Research of the National Institute of Standards and Technology* 99(4):327.

Shibata, T. 1996. 1996 W.R. Whitney Award lecture: Statistical and stochastic approaches to localized corrosion. *Corrosion* 52(11):813-830.

Shibata, T. 2013. Birth and death stochastic process in pitting corrosion and stress corrosion cracking. *ECS Transactions* 50(31):13.

Shibata, T., and T. Takeyama. 1976. Pitting corrosion as a stochastic process. *Nature* 260(5549):315-316.

Shibata, T., and T. Takeyama. 1977. Stochastic theory of pitting corrosion. *Corrosion* 33(7):243-251.

Shiu, H. Y., and R. W. Cheung. 2008. Long-term durability of steel soil nails in Hong Kong. *HKIE Transactions* 15(3):24-32.

Shreir, L. L., R. A. Jarman, and G. T. Burstein. 1994. *Corrosion*, vol. 1, 3rd ed. Oxford, UK: Butterworth Heinemann Ltd.

Sidik, W. S., H. Canakci, I. H. Kilic, and F. Celik. 2014. Applicability of biocement for organic soil and its effect on permeability. *Geomechanics & Engineering* 7(6):649-663.

Singh, R. P. 2014. *Pipeline integrity handbook*. New York: Elsevier.

Snyder, G., B. Parker, D. McEwen, R. Donnelly, and R. Murray. 2007. *Assessing the reliability of existing anchor installations at Loch Alva and Log Falls dams*. Paper presented at CDA Annual Conference, Markham, Ontario, Canada.

Soomro, A. A., A. A. Mokhtar, J. C. Kurnia, N. Lashari, H. Lu, and C. Sambo. 2022. Integrity assessment of corroded oil and gas pipelines using machine learning: A systematic review. *Engineering Failure Analysis* 131:105810.

Soong, J. L., C. L. Phillips, C. Ledna, C. D. Koven, and M. S. Torn. 2020. CMIP5 models predict rapid and deep soil warming over the 21st century. *Journal of Geophysical Research: Biogeosciences* 125(2):e2019JG005266.

Spark, A., K. Wang, I. Cole, D. Law, and L. Ward. 2020. Microbiologically influenced corrosion: A review of the studies conducted on buried pipelines. *Corrosion Reviews* 38(3):231-262.

Statfull, R. F., and C. Seim. 1979. *Corrosion of steel piles at the Richmond-San Rafael bridge*. FHWA-CA-78-1. Sacramento, CA: California Department of Transportation.

Stefanescu, D. M. 1990. Classification and basic metallurgy of cast iron. In *ASM handbook, volume 1: Properties and selection: Irons, steels, and high-performance alloys*, edited by ASM Handbook Committee. Materials Park, OH: ASM International. Pp. 3-11.

Stern, M., and A. L. Geary. 1957. Electrochemical polarization: I. A theoretical analysis of the shape of polarization curves. *Journal of the Electrochemical Society* 104(1):56.

Stevenson, A., J. A. Cray, J. P. Williams, R. Santos, R. Sahay, N. Neuenkirchen, C. D. McClure, I. R. Grant, J. D. Houghton, and J. P. Quinn. 2015. Is there a common water-activity limit for the three domains of life? *ISME Journal* 9(6):1333-1351.

Steward, D. R. 2020. *Analytic element method: Complex interactions of boundaries and interfaces*. Oxford, UK: Oxford University Press.

Sundholm, S. 1987. The quality control of rock bolts. In *Proceedings of the 6th International Congress on Rock Mechanics, August, 1987, Montreal, Canada*. Rotterdam and Boston: A.A. Balkema.

Taamneh, M., and R. Liang. 2010. Update on long-term monitoring results at ATB-90. In *TRB 89th Annual Meeting compendium of papers* [DVD]. Washington, DC: Transportation Research Board.

Tayama, S., Y. Kawai, and H. Maeno. 1996. An investigation of the durability of the soil nailing method. In *Proceedings of the International Symposium on Earth Reinforcement*, edited by A. A. Kyushu. Rotterdam, Netherlands: CRC Press/Balkema. Pp. 161-166.

Tex-620-J. 2002. *Determining chloride and sulfate content in soils*. Austin, TX: Texas Department of Transportation.

Tex-620-M. 2018. *Test procedure for determining the conductivity, pH, sulfate content, and chloride content of soil and coarse aggregate*. Austin, TX: Texas Department of Transportation.

Thapalia, A., D. M. Borrok, S. Nazarian, and J. Garibay. 2011. Assessment of corrosion potential of coarse backfill aggregates for mechanically stabilized earth walls. *Transportation Research Record* 2253(1):63-72.

Timmerman, D. 1992. *Evaluation of mechanically stabilized embankments as support for bridge structures. Final report.* FHWA/OH-91/014. Washington, DC: Federal Highway Administration.

Törnqvist, J., and J. Lehtonen. 1999. Estimation of corrosion rate in durability design of steel piles. In *Proceedings, 12th European Conference on Soil Mechanics and Foundation Engineering, Amsterdam*. Pp. 433-438.

True, F. W. 1913. *A history of the first half-century of the National Academy of Sciences, 1863–1913*. Baltimore, MD: Lord Baltimore Press.

Truffert, C., J. Gance, O. Leite, and B. Texier. 2019. *New instrumentation for large 3D electrical resistivity tomography and induced polarization surveys*. Paper presented at International Workshop on Gravity, Electrical & Magnetic Methods and Their Applications, Xi'an, China.

Tucker, S. E., J. L. Briaud, S. Hurlebaus, M. E. Everett, and R. Arjwech. 2015. Electrical resistivity and induced polarization imaging for unknown bridge foundations. *Journal of Geotechnical and Geoenvironmental Engineering* 141(5).

Uhlig, H. H., and R. W. Revie. 1985. *Corrosion and corrosion control*, 3rd ed. New York: John Wiley & Sons.

Umar, M., K. A. Kassim, K. Tiong, and P. Chiet. 2016. Biological process of soil improvement in civil engineering: A review. *Journal of Rock Mechanics and Geotechnical Engineering* 8(5):767-774.

University at Buffalo. 2022. About MCEER. https://www.buffalo.edu/mceer/about.html (accessed January 24, 2023).

Valor, A., F. Caleyo, L. Alfonso, D. Rivas, and J. Hallen. 2007. Stochastic modeling of pitting corrosion: A new model for initiation and growth of multiple corrosion pits. *Corrosion Science* 49(2):559-579.

Van Nostrand, R. G., and K. L. Cook. 1966. *Interpretation of resistivity data*. Washington, DC: U.S. Government Printing Office.

Vazquez, S. 2014. *Corrosion behavior of galvanized steel reinforcements in MSE walls in the presence of soil organics*. M.Sc. thesis. University of British Columbia.

Velázquez, J. C., F. Caleyo, A. Valor, and J. M. Hallen. 2009. Predictive model for pitting corrosion in buried oil and gas pipelines. *Corrosion* 65(5):332-342.

Velázquez, J. C., F. Caleyo, A. Valor, and J. M. Hallen. 2010. Field study—pitting corrosion of underground pipelines related to local soil and pipe characteristics. *Corrosion* 66(1):016001.

Venzlaff, H., D. Enning, J. Srinivasan, K. J. Mayrhofer, A. W. Hassel, F. Widdel, and M. Stratmann. 2013. Accelerated cathodic reaction in microbial corrosion of iron due to direct electron uptake by sulfate-reducing bacteria. *Corrosion Science* 66:88-96.

Vigneron, A., E. B. Alsop, B. Chambers, B. P. Lomans, I. M. Head, and N. Tsesmetzis. 2016. Complementary microorganisms in highly corrosive biofilms from an offshore oil production facility. *Applied and Environmental Microbiology* 82(8):2545-2554.

Vigneron, A., P. Cruaud, E. Alsop, J. R. de Rezende, I. M. Head, and N. Tsesmetzis. 2018. Beyond the tip of the iceberg: A new view of the diversity of sulfite- and sulfate-reducing microorganisms. *ISME Journal* 12(8):2096-2099.

Vilda, W. S., III. 2009. Design Manual for Roads and Bridges. BD 42/00, vol. 2, Section 1, Part 2.

Volk, E., S. C. Iden, A. Furman, W. Durner, and R. Rosenzweig. 2016. Biofilm effect on soil hydraulic properties: Experimental investigation using soil-grown real biofilm. *Water Resources Research* 52(8):5813-5828.

Wang, C., M. A. Briggs, F. D. Day-Lewis, and L. D. Slater. 2021. Characterizing physical properties of streambed interface sediments using in situ complex electrical conductivity measurements. *Water Resources Research* 57(2):e2020WR027995.

Wang, H., A. Yajima, R. Y. Liang, and H. Castaneda. 2015a. A Bayesian model framework for calibrating ultrasonic in-line inspection data and estimating actual external corrosion depth in buried pipeline utilizing a clustering technique. *Structural Safety* 54:19-31.

Wang, H., A. Yajima, R. Y. Liang, and H. Castaneda. 2015b. A clustering approach for assessing external corrosion in a buried pipeline based on hidden Markov random field model. *Structural Safety* 56:18-29.

Wang, H., A. Yajima, R. Y. Liang, and H. Castaneda. 2015c. Bayesian modeling of external corrosion in underground pipelines based on the integration of Markov chain Monte Carlo techniques and clustered inspection data. *Computer-Aided Civil and Infrastructure Engineering* 30(4):300-316.

Wang, H., A. Yajima, R. Y. Liang, and H. Castaneda. 2016. Reliability-based temporal and spatial maintenance strategy for integrity management of corroded underground pipelines. *Structure and Infrastructure Engineering* 12(10):1281-1294.

Wang, H., A. Yajima, and H. Castaneda. 2019. A stochastic defect growth model for reliability assessment of corroded underground pipelines. *Process Safety and Environmental Protection* 123:179-189.

Wang, H., S. Sreeharan, and H. Castaneda. 2021. *Mapping indication severity using Bayesian machine learning from inspection data by considering the impact of soil corrosivity.* Paper presented at CORROSION 2021, Virtual, April 1. Paper Number NACE-2021-16749.

Wang, X., H. Wang, F. Tang, H. Castaneda, and R. Liang. 2021. Statistical analysis of spatial distribution of external corrosion defects in buried pipelines using a multivariate Poisson-lognormal model. *Structure and Infrastructure Engineering* 17(6):741-756.

Washko, S. D., and G. Aggen. 1990. Wrought stainless steels. In *ASM handbook, volume 1: Properties and selection: Irons, steels, and high-performance alloys*, edited by ASM Handbook Committee. Materials Park, OH: ASM International. Pp. 841-907.

Weisberg, S. 2005. *Applied linear regression*, vol. 528. New York: John Wiley & Sons.

Wen, Y., C. Cai, X. Liu, J. Pei, X. Zhu, and T. Xiao. 2009. Corrosion rate prediction of 3C steel under different seawater environment by using support vector regression. *Corrosion Science* 51(2):349-355.

Whiffin, V. S., L. A. van Paassen, and M. P. Harkes. 2007. Microbial carbonate precipitation as a soil improvement technique. *Geomicrobiology Journal* 24(5):417-423.

Wilkinson, M. D., M. Dumontier, I. J. Aalbersberg, G. Appleton, M. Axton, A. Baak, N. Blomberg, J.-W. Boiten, L. B. da Silva Santos, and P. E. Bourne. 2016. The FAIR guiding principles for scientific data management and stewardship. *Scientific Data* 3(1):1-9.

WisDOT (Wisconsin Department of Transportation). 2015. *I-43 Leo Frigo Memorial Bridge—investigation report.* Madison, WI.

Withiam, J. L., K. L. Fishman, and M. P. Gaus. 2002. *NCHRP Report 477: Recommended practice for evaluation of metal-tensioned systems in geotechnical applications.* Washington, DC: Transportation Research Board.

Wong, I. H., and K. H. Law. 1999. Corrosion of steel H piles in decomposed granite. *Journal of Geotechnical and Geoenvironmental Engineering* 125(6):529-532.

Yajima, A., R. Liang, H. Rivera, L. Martinez, A. Karayan, and H. Castaneda. 2014. *The application of macro modeling concept for the soil/coating external corrosion for ECDA process by using statistical tools.* Paper presented at CORROSION 2014, San Antonio, Texas, March. Paper Number NACE-2014-4412.

Yajima, A., H. Wang, R. Y. Liang, and H. Castaneda. 2015. A clustering based method to evaluate soil corrosivity for pipeline external integrity management. *International Journal of Pressure Vessels and Piping* 126:37-47.

Yu, T., and P. L. Bishop. 2001. Stratification and oxidation—reduction potential change in an aerobic and sulfate-reducing biofilm studied using microelectrodes. *Water Environment Research* 73(3):368-373.

Zhang, S., and W. Zhou. 2015. Probabilistic characterisation of metal-loss corrosion growth on underground pipelines based on geometric Brownian motion process. *Structure and Infrastructure Engineering* 11(2):238-252.

Zhang, S., W. Zhou, and H. Qin. 2013. Inverse Gaussian process-based corrosion growth model for energy pipelines considering the sizing error in inspection data. *Corrosion Science* 73:309-320.

Zhang, Y., B. M. Ayyub, and J. F. Fung. 2022. Projections of corrosion and deterioration of infrastructure in United States coasts under a changing climate. *Resilient Cities and Structures* 1(1):98-109.

Zhao, X., P. Gong, G. Qiao, J. Lu, X. Lv, and J. Ou. 2011. Brillouin corrosion expansion sensors for steel reinforced concrete structures using a fiber optic coil winding method. *Sensors* 11(11):10798-10819.

Zhou, E., J. Wang, M. Moradi, H. Li, D. Xu, Y. Lou, J. Luo, L. Li, Y. Wang, and Z. Yang. 2020. Methanogenic archaea and sulfate reducing bacteria induce severe corrosion of steel pipelines after hydrostatic testing. *Journal of Materials Science & Technology* 48:72-83.

Zhou, W. 2010. System reliability of corroding pipelines. *International Journal of Pressure Vessels and Piping* 87(10):587-595.

Zhou, W., C. Gong, and H. Hong. 2017. New perspective on application of first-order reliability method for estimating system reliability. *Journal of Engineering Mechanics* 143(9):04017074.

Appendix A

Biographical Sketches of Committee Members

Scott A. Anderson (*Chair*) is a principal geotechnical engineer at BGC Engineering in Golden, Colorado. He has more than 35 years of geotechnical design and construction experience in the transportation, water resources, mining, and pipeline industries. In addition to working as a consultant to these industries, Dr. Anderson is a former researcher, professor, and government agency leader. He has led and provided oversight and review to geohazard and construction site characterizations, and design and construction activities, often focusing on risk-based assessments. He is devoted to learning and advancement through the research and deployment of new technology in many areas of practice as demonstrated through work as a committee chair and member of several technical committees of the National Academies of Sciences, Engineering, and Medicine's Transportation Research Board (TRB) and support to TRB's National Cooperative Highway Research Program for nearly 20 years. He has been a Steering Committee member for the Geotechnical Extreme Event Reconnaissance Association for more than 10 years and is a member of American Society of Civil Engineering Geo-Institute committees, including the Innovative Technologies and Tools in Geotechnical Engineering Committee. Prior to joining BGC Engineering, Dr. Anderson was the geotechnical services team leader for the Federal Highway Administration (FHWA) Resource Center from 2008 to 2017, and he held geotechnical leadership roles for the Federal Lands Highway Division of FHWA for 6 years. Selected awards include the FHWA Engineer of the Year in 2014, the K.B. Woods Award in 2016 from TRB, and the 2017 Jahns Distinguished Lecturer for the Association of Engineering Geologists and the Geological Society of America. Dr. Anderson holds a B.A. and an M.S. in engineering geology from the University of Colorado Boulder and Colorado State University, respectively. He received an M.S. and a Ph.D. in civil engineering from the University of California, Berkeley.

Mersedeh Akhoondan has more than 15 years of experience in condition assessment and rehabilitation of metallic and reinforced concrete infrastructures, specializing in service-life modeling of different types of infrastructure. She has designed and managed corrosion testing programs, developed rehabilitation plans and specifications for corrosion control systems, performed failure analysis and metallographic evaluations of field coupons, and conducted soil corrosivity studies. Dr. Akhoondan is a registered civil engineer in the states of Florida, New York, and Texas. She has authored or co-authored more than 30 technical publications. Dr. Akhoondan received her B.S. and M.S. in civil engineering and a Ph.D. in corrosion engineering from the University of South Florida.

Susan E. Burns is the Dwight H. Evans Professor and the associate chair for administration and finance in the School of Civil and Environmental Engineering at the Georgia Institute of Technology. Dr. Burns's research focuses on applications in geoenvironmental engineering with particular emphasis on beneficial use of waste materials; erosion, infiltration, and stormwater treatment on roadway rights-of-way; bio-mediated ground improvement; and fundamental chemical and engineering behavior of soils. Funding for her research group has come from a range of agencies, including the National Science Foundation (NSF), the U.S. Department of Energy, the U.S. Army Corps of Engineers, the U.S. Department of Education, the Virginia Transportation Research Council, the Georgia Department of Transportation, Southern Company, and other industrial sources. Dr. Burns is a recipient of the NSF CAREER Award and a fellow of the American Society of Civil Engineers, and she was named the 2020 Engineer of the Year by the Georgia Society of Professional Engineers. In 2021, Dr. Burns received the Class of 1940 W. Howard Ector Outstanding Teacher Award, which is Georgia Tech's highest award for teaching, and was named one of the 100 most influential women in Georgia Engineering by the American Council of Engineering Companies of Georgia. Dr. Burns earned a B.C.E. (1990) and an M.S. in civil engineering (geotechnical) (1996), an M.S. in environmental engineering (1996), and a Ph.D. in civil engineering (1997), all from Georgia Tech.

Homero Castaneda-Lopez is an associate professor and the director at the National Corrosion and Materials Reliability Laboratory within Texas A&M University (TAMU). He is also the instructor for NACE Cathodic Protection certifications. Dr. Castaneda has 19 years of experience using electrochemical and nondestructive techniques to monitor interfacial phenomena in materials and theoretical modeling of corrosion science and engineering, energy generation and storage, and infrastructure and electrochemical processes for different industries. He has been the principal investigator for multiple projects on corrosion science and engineering for the U.S. Department of Energy, the U.S. Department of Defense, the U.S. Department of Transportation, and several Fortune 500 companies. Before joining TAMU, he worked for 5 years at the University of Akron (2011 to 2015) as an assistant professor and before that at the Battelle Memorial Institute as a senior scientist (2006–2010) in advanced materials and energy systems in Columbus, Ohio. Before Battelle, he was the technical director and research leader of corrosion, materials, and pipelines in the Mexican Petroleum Institute for 3 years. He has authored more than 100 peer-reviewed papers in the areas of corrosion science and engineering, coatings degradation and materials reliability, materials characterization, and electrochemical impedance spectroscopy. He holds eight patents and four copyrights. He received the H.H. Uhlig award from NACE International in 2018. He is NACE Fellow Class of 2019. He is the editor of three journals related to electrochemistry, corrosion, and pipelines. He received his B.S. in chemical metallurgical engineering (1994) and M.S. in materials science (1997) from the National University of México and his Ph.D. in materials science and engineering from Penn State University in 2001.

Kenneth L. Fishman is a principal at McMahon and Mann Consulting Engineering and Geology, P.C. and is the leader of its Earth Reinforcement Testing Division. He has 40 years of combined experience in civil and geotechnical engineering that includes teaching, research, and consulting and is an expert on state-of-the-art techniques for performance monitoring, characterization of corrosion potential, and service-life modeling for mechanically stabilized earth (MSE) walls and other geotechnical applications. He has been a consultant, co-principal investigator (PI), or PI on various projects sponsored by the National Cooperative Highway Research Program (NCHRP), the Federal Highway Administration (FHWA), the Multidisciplinary Center for Earthquake Engineering Research, and various state departments of transportation (DOTs) on projects related to service-life design, condition assessment, corrosion monitoring, and durability studies for buried steel including MSE structures, elements of deep foundation systems, and pipelines. These projects include research, implementations of research results, training, and applications of the state-of-the-art techniques for performance monitoring and characterization of corrosion potential. He is author of more than 50 publications in the area of geotechnical engineering, including FHWA-HRT-05-067, NCHRP Report 477, FHWA-NHI-09-087, NCHRP Report 675, NCHRP Research Report 958, several comprehensive reports describing results from FHWA/state DOT-sponsored demonstration projects, and pooled-fund studies. He has participated in numerous workshops and webinars sponsored by NCHRP, FHWA, the American Association of State Highway and Transportation Officials (AASHTO), the Transportation Research Board (TRB), and others including two recent webinars related to asset management and performance modeling sponsored by the FHWA Office of Asset Management and Resource Center, AASHTO, and TRB. Dr. Fishman earned his Ph.D. for his work in geotechnical engineering from the University of Arizona in 1988.

Gerald S. Frankel is a distinguished professor of engineering, a professor of materials science and engineering, and the director of the Fontana Corrosion Center at The Ohio State University (OSU). Prior to joining OSU in 1995, he was a postdoctoral researcher at the Swiss Federal Technical Institute in Zurich and then a research staff member at the IBM Watson Research Center in Yorktown Heights, New York. His primary research interests are in the passivation and localized corrosion of metals and alloys, corrosion inhibition, protective coatings, and atmospheric corrosion. He has authored more than 300 papers in peer-reviewed journals. He is a member of the editorial board of *Corrosion* and a fellow of the Association for Materials Protection and Performance, the Electrochemical Society, and ASM International. He received the W.R. Whitney Award from NACE International in 2015, the U.R. Evans Award from the UK Institute of Corrosion in 2011, the OSU Distinguished Scholar Award in 2010, the 2010 Electrochemical Society's Olin Palladium Award, and the Alexander von Humboldt Foundation Research Award for Senior U.S. Scientists in 2004. From 2012 to 2016, he served as a member of the Nuclear Waste Technical Review Board after being appointed by President Obama. In 2016, he became the director of a U.S. Department of Energy–funded Engineering Frontier Research Center focused on the performance of nuclear waste forms. He earned his Sc.B. in materials science engineering from Brown University and his Sc.D. in materials science and engineering from the Massachusetts Institute of Technology.

Stacey Kulesza is an associate professor in civil engineering at Texas State University where her research focuses on transportation geotechnical infrastructure design, performance-monitoring, and durability studies with particular emphasis on the electrical properties of soils; integrated site characterization; anthropogenic impacts on soil properties; and soil erosion potential. She also studies asset-based frameworks toward creating authentic engineering identities, particularly in military veterans and women. She currently serves on two technical committees of the Transportation Research Board and the American Society of Civil Engineers (ASCE) Geo-Institute including serving as the chair for the ASCE Geo-Institute Outreach and Engagement Committee. She is a registered professional engineer in the state of Kansas and was the 2020 Tri-Valley Young Engineer of the Year in Kansas. She received her B.S., M.E., and Ph.D. in civil engineering with a focus in geotechnical engineering from Texas A&M University.

Brenda J. Little retired in January 2018 from the Naval Research Laboratory, Stennis Space Center, where she served as a senior scientist. Her career has focused on the investigation of microorganism–material interactions including biodeterioration, biodegradation, and bioremediation (i.e., chemistries produced by microorganisms). Her publications include two co-authored books and more than 100 peer-reviewed journal articles on these topics. Dr. Little is now an independent consultant and the sole proprietor of B.J. Little Corrosion Consulting, LLC, and she serves as the past president of the International Biodeterioration and Biodegradation Society (IBBS). Dr. Little is a fellow of NACE International and on the editorial board for *International Biodeterioration and Biodegradation*, the official journal for the IBBS. Dr. Little received her Ph.D. in chemistry from Tulane University.

Randall W. Poston (NAE) is a senior consultant at Pivot Engineers, a structural engineering consulting firm in Austin, Texas. He was a Neil Armstrong Distinguished Visiting Fellow at the Purdue University College of Engineering from 2019 to 2021. For the past 35 years, Dr. Poston has been engaged in the evaluation, repair, strengthening, and design of more than 700 structures. His expertise includes the investigation of structural failures, evaluation of corrosion of steel in concrete, structural concrete repair and strengthening design, and nondestructive testing of concrete structures. He has been elected a fellow of the American Concrete Institute (ACI), the American Society of Civil Engineers, the Post-Tensioning Institute, and the International Association for Bridge and Structural Engineering; is an active member of numerous national and international technical committees including being a current member of ACI Committee 318; and was the chair of Committee 318 during the 2014 code cycle. Dr. Poston was the past president of ACI from 2019 to 2020. Dr. Poston was elected to the U.S. National Academy of Engineering in 2017. He received his B.S., M.S., and Ph.D. in civil engineering from The University of Texas at Austin.

Elizabeth Rutherford is a senior metallurgical engineer at Energy Transfer with more than 15 years of experience in pipeline integrity, failure investigation, root-cause analysis, and quality assurance. She has worked on pipeline failures in a variety of terrains in the United States in addition to numerous preventive inspections and repairs. She is actively involved in both Pipeline Research Council International and joint industry projects addressing research gaps and identifying technologies to move the industry forward safely. For the past 3 years, her focus has been on the quality assurance side of the business, monitoring the acquisition and production of pipe for new construction projects. Prior to joining Energy Transfer, Ms. Rutherford spent 3 years with the Nuclear Regulatory Commission as a reactor inspector with special emphasis on in-service inspection. Ms. Rutherford received her B.S. in metallurgical engineering from the University of Missouri–Rolla (now the Missouri University of Science and Technology).

Appendix B

Meeting and Workshop Agendas

Committee on Corrosion of Buried Steel at New and In-Service Infrastructure
Board on Earth Sciences and Resources
National Academies of Sciences, Engineering, and Medicine
Meeting 1, August 7, 2020

OPEN SESSION
11:00 a.m.–12:45 p.m. EDT
Sponsor Input

Meeting Objectives:
- Learn how the sponsors are interpreting the Statement of Task and what types of recommendations will be most helpful.
- Review any discrepancies among sponsors and come to some agreement regarding how to bound the study.

11:00   **Introductions, description of session objectives**
Scott Anderson, Chair, COGGE

11:05   **Sponsor input**
Jennifer Nicks and Silas Nichols, Federal Highway Administration
Vanessa Bateman and Matt Smith, U.S. Army Corps of Engineers
Brad Keelor, Geo-Institutes of the American Society of Civil Engineers
Peggy Hagerty Duffy, ADSC

12:05   **Q&A with sponsors**
Are sponsors and committee in agreement on study definitions and study boundaries?

11:45   **Break**

12:45   **Adjourn**

Laboratory and Field Geotechnical Characterization for Improved Steel Corrosion Modeling

March 9–10, 2021
11:00 a.m.–6:00 p.m. (EST)
Virtual Workshop Agenda

The National Academies' Committee on the Corrosion of Buried Steel at New and In-Service Infrastructure is hosting a 2-day virtual workshop to gather information on field, laboratory, and modeling methods for characterizing corrosion of steel buried in earth materials and new developments in the prediction and monitoring of corrosion of steel in earth applications and environments. These sessions are part of the National Academies' consensus study examining the state of knowledge and technical issues regarding the corrosion of steel used for each application and identifying knowledge gaps and research needed to improve long-term performance of steel.

DAY 1
SESSION 1—Modeling
11:00 a.m.–2:15 p.m. EST
Identifying corrosion modeling approaches for improved understanding of corrosion potential (E_{corr}), rates, assessment, and management

Session Objectives:
- Define common vocabulary for workshop participants.
- Describe current deterministic modeling approaches and their limitations.
- Identify emerging (nondeterministic) modeling approaches and challenges to their application.

Prompting Questions for Speakers:
- What do current deterministic models deliver, and with what kinds of limitations and uncertainties?
- Are there assumptions in deterministic corrosion models that are no longer needed or valid?
- What are the opportunities for improving deterministic models and what steps are needed to do so?
- What limitations and uncertainties in deterministic models can be addressed with nondeterministic models and what steps are needed to do so?

Prerecorded presentation for prior viewing: Predicting Corrosion in Soils for Infrastructure Applications: A Review and Recent Developments, *Rob Melchers, University of Newcastle*

11:00   **Welcome, introductions, discussion of Statement of Task, and workshop objectives**
         Scott Anderson, BGC Engineering Inc., Committee Chair

11:30   **Presentation: Determinism in science and engineering**
         Digby Macdonald, University of California, Berkeley

12:00   **Discussion with Digby Macdonald**
         Moderator: Homero Castaneda, Texas A&M University

12:30   **Panel discussion on current approaches to modeling corrosion of steel in earth materials**
         Moderator: Homero Castaneda, Texas A&M University

Five-minute presentations by each panelist
Greg Baecher, University of Maryland
Han-Ping Hong, Western University
Mark Orazem, University of Florida
Alberto Sagues, University of South Florida
Hui (Jack) Wang, University of Dayton

12:55 **Panel discussion moderated by Homero Castaneda including the following questions:**

- What assumptions go into pre- and post-construction modeling and are those assumptions correct?
- What types of information (i.e., model inputs) do modelers utilize?
- In terms of model performance, where do models predict well, and where do they break down?
- What are the outputs of and uncertainties associated with deterministic versus nondeterministic models?

2:15 **Break**

SESSION 2—Laboratory
3:00 p.m.–6:00 p.m. EST
Laboratory measurements as model parameters: Past and future

Session Objectives:
- Identify laboratory data that are supplied and used in modeling.
- Describe the limitations of laboratory methods for characterizing corrosivity observed in the field.
- Identify differences between laboratory data that are available versus data desired to better inform model inputs.
- Identify how laboratory practices might evolve to better meet needs for models.

Prompts for Speakers:
- With respect to characterizing corrosion rate of buried steel, describe (a) current laboratory practices, (b) opportunities for and evolution of practice to better meet needs, and (c) barriers to evolution for the following applications (as appropriate):
    1. Characterizing spatial variability
    2. Characterizing environmental variability
    3. Replicating the as-built environment
    4. Providing parameters for model input

3:00 **Description of session and session goals**
Scott Anderson, BGC Engineering Inc., Committee Chair

3:05 **Brief presentations: Setting the stage for discussion in response to above prompts**

What do model-environments encounter on underground pipelines
John Beavers, DNV GL USA, Inc.

3:20 **Laboratory testing for internal and external corrosion of buried pipelines?**
Frank Cheng, University of Calgary

3:35 **Laboratory practices for determination of electrochemical properties of soil and rock and method limitations**
Karl Fletcher, Bowser-Morner, Inc.

3:50   **Field investigation of corrosion of buried metallic reinforcement behind retaining walls**
Bob Parsons, University of Kansas

4:05   **Instruction for breakout discussions and transition time**
Scott Anderson

4:10   **Breakout discussions**

Breakout Room Theme	#1: Characterizing spatial variability	#2: Characterizing environmental variability	#3: Replicating the as-built environment	#4: Providing parameters for model input	#5: Providing parameters for model input

Each breakout room to discuss the following:

- What are current laboratory practices with respect to the breakout theme?
- What are the opportunities to better meet needs with respect to the breakout theme?
- What are the barriers to making the above improvements?

5:00   **Transition to plenary**

5:10   **Plenary session: Summaries from breakout sessions**
Moderator: Scott Anderson, BGC Engineering Inc., Committee Chair

6:00   **Adjourn Day 1**

DAY 2
SESSION 3—Field
11:00 a.m.–1:45 p.m. EST
Informing modeling through field characterization and performance monitoring data

Session Objectives:
- Learn about current and emerging field methodologies and their uncertainties.
- Understand the utility of field data (e.g., modeling as direct input, as supplemental guide, or monitoring data).
- Consider if and how field data uncertainties are represented in corrosion models.

Prerecorded presentation for prior viewing: Fundamentals in Geophysics for Assessing Corrosion
Mark Everett, Texas A&M University

Prerecorded panelist intros and response to questions for prior viewing:
- How do we get valuable in situ field data regarding the extents, rates, and mechanisms of corrosion? *Khalid Farrag, Gas Technology Institute*
- In what ways are field tests and measurements relevant with respect to managing buried steel performance? *Kathryn Griswell, California Department of Transportation*
- What measurement technologies related to measuring corrosion of steel in reinforced concrete might be used in geotechnical applications? *Amir Poursaee, Clemson University*
- How are, and can, field data uncertainties be represented in corrosion models? *Naresh Samtani, NCS GeoResources*

11:00   **Welcome, objectives of Day 2**
Scott Anderson, Committee Chair

11:20   **Five-minute recap of prerecorded presentation**
Mark Everett, Texas A&M University

11:25   **Clarifying questions from committee for Mark Everett**
Moderator: Stacey Kulesza, Kansas State University

11:35   **Presentation: State of field monitoring and characterization for estimating corrosion-related indicators and parameters**
Soheil Nazarian, The University of Texas at El Paso

11:55   **Clarifying questions from committee for Soheil Nazarian**
Moderator: Stacey Kulesza, Kansas State University

12:05   **Presentation: Field characterization of dynamic hydrogeologic, geochemical and microbial conditions that affect corrosion**
Jennifer McIntosh, University of Arizona

12:25   **Clarifying questions from committee for Jennifer McIntosh**
Moderator: Stacey Kulesza, Kansas State University

12:35   **Intro for panelists and panel discussion: State of practice for field monitoring across different industries: Current limitations and uncertainties in assumptions**
Moderator: Stacey Kulesza, Kansas State University
Khalid Farrag, Gas Technology Institute
Kathryn Griswell, California Department of Transportation
Amir Poursaee, Clemson University
Naresh Samtani, NCS GeoResources
Mark Everett, Texas A&M University
Soheil Nazarian, The University of Texas at El Paso
Jennifer McIntosh, University of Arizona

1:35   **Summarize important points from session**
Ken Fishman, McMahon and Mann Consulting Engineers

1:45   **Break**

SESSION 4—The Future
2:45 p.m.–5:25 p.m. EST
Future modeling and laboratory and field geotechnical characterization
for improved understanding of buried steel corrosion

Session Objectives:
- Explore how existing lab and field techniques might be refined to improve modeling.
- Explore how to refine modeling to better constrain uncertainties.

Prompting Questions for Discussion:
- How might work practices for characterization of earth materials or for long-term management of buried steel be modified based on what has been discussed during this workshop?
- What developments in modeling, laboratory, or field work might lead to improved understanding of corrosion potential, rates, assessment, and management?

2:45   **Introduction to session objectives**
        Scott Anderson, BGC Engineering Inc., Committee Chair

2:50   **Flash talks on emerging approaches to improve corrosion modeling**
        Moderator: Scott Anderson, BGC Engineering Inc., Committee Chair
        Arturo Bronson, The University of Texas at El Paso
        Han-Ping Hong, Western University
        Burkan Isgor, Oregon State University
        Erik Loehr, University of Missouri
        Wenxing Zhou, Western University

3:20   **Discussion with flash talk speakers**
        Moderator: Ken Fishman, McMahon and Mann Consulting Engineers

4:20   **Summary discussion from previous sessions**
        Moderator: Susan Burns, Georgia Institute of Technology
            Session 1 take home messages, *Homero Castaneda, Texas A&M University*
            Session 2 take home messages, *Scott Anderson, BGC Engineering Inc., Committee Chair*
            Session 3 take home messages, *Stacey Kulesza, Kansas State University*

5:05   **Open mic: What would you like to say that you haven't heard at the workshop?**
        Moderator: Scott Anderson, BGC Engineering Inc., Committee Chair

5:20   **Closing remarks**
        Scott Anderson, BGC Engineering Inc., Committee Chair

5:25   **Workshop adjourns**

Invited Speakers

Greg Baecher, University of Maryland
John Beavers, DNV GL
Arturo Bronson, The University of Texas at El Paso
Frank Cheng, University of Calgary
Mark Everett, Texas A&M University
Khalid Farrag, Gas Technology Institute
Karl Fletcher, Bowser Morner
Kathryn Griswell, California Department of Transportation
Han-Ping Hong, Western University
Burkan Isgor, Oregon State University
Erik Loehr, University of Missouri
Digby Macdonald, University of California, Berkeley
Jennifer McIntosh, University of Arizona
Robert Melchers, University of Newcastle, Australia

Soheil Nazarian, The University of Texas at El Paso
Mark Orazem, University of Florida
Bob Parsons, University of Kansas
Amir Poursaee, Clemson University
Naresh Samtani, NCS GeoResources, LLC
Alberto Sagues, University of South Florida
Hui (Jack) Wang, University of Dayton
Wenxing Zhou, Western University

Participants

Mersedeh Akhoondan, HDR Engineering
Christopher Alexander, University of South Florida
Peter Anderson, Reinforced Earth
Scott A. Anderson, BGC Engineering
Hans Arlt, U.S. Nuclear Regulatory Commission
Aziz Asphahani, QuesTek Innovations
Recep Avci, Montana State University
Vanessa C. Bateman, U.S. Army Corps of Engineers
Christine Beyzaei, Exponent
Jon Bischoff, Utah Department of Transportation
Giovanna Biscontin, National Science Foundation
Keith Brabant, Reinforced Earth
Michael Carey Brown, WSP USA
Susan E. Burns, Georgia Institute of Technology
Leonardo Caseres, Southwest Research Institute
Homero Castaneda, National Corrosion and Materials Reliability Laboratory
Craig Davis, Los Angeles Department of Water and Power (retired)
Jerry DiMaggio, Applied Research Associates
Peggy Hagerty Duffy, ADSC
James Ellor, Elzly Technology Corporation
Ray Fassett, Condon-Johnson & Associates
Kenneth L. Fishman, McMahon and Mann Consulting Engineers P.C.
Gerald S. Frankel, The Ohio State University
Marcus Galvan, Foresight Planning and Engineering Service
Robert Gladstone, Association for Mechanically Stabilized Earth
Mike Gomez, University of Washington
Anand Govindasamy, Geosyntec Boston
David Harris, Integral Engineering Co.
Tom Hayden, Engineering Director Inc.
Harold Hilfiker, Hilfiker Retaining Walls
Terry Holman, Geosyntec Consultants
Reggie Holt, U.S. Department of Transportation
Navid Jafari, Louisiana State University
Leszek Janusz, ViaCon Polska
Brad Keelor, Geo-Institute of the American Society of Civil Engineers
Stacey Kulesza, Kansas State University
Kingsley Lau, Florida International University
Brenda J. Little, B.J. Little Corrosion Consulting LLC
Daryl Little, Bureau of Reclamation

Allen Marr, Geocomp
Robert D. Moser, U.S. Army Corps of Engineers
Silas Nichols, U.S. Department of Transportation
Jennifer Nicks, U.S. Department of Transportation
Justin Ocel, U.S. Department of Transportation
Larry Olson, Olson Engineering Inc.
Joy Pauschke, National Science Foundation
Randall Poston, Pivot Engineers
Kyle Rollins, Brigham Young University
Elizabeth Rutherford, Energy Transfer
Tom Schwerdt, Texas Department of Transportation
Jeff Segar, Braun Intertec Corporation
John Senko, University of Akron
Preet Singh, Georgia Institute of Technology
Matthew D. Smith, U.S. Army Corps of Engineers
Derek Soden, U.S. Department of Transportation
Pete Speier, Williams Form Engineering Corp
Narasi Sridhar, MC Consult LLC
Elizabeth Trillo, Southwest Research Institute
Joseph Turk, Tennessee Valley Authority
Leon van Paassen, Arizona State University
Mark Vessely, BGC Engineering
Kevin White, E.L. Robinson Engineering
John Wolodko, The University of Alberta
Hui Yu, TRC Companies Inc.

Appendix C

Acronyms and Abbreviations

AASHTO	American Association of State Highway and Transportation Officials
AC	alternating current
ADSC	International Association of Foundation Drilling
AMPP	Association for Materials Protection and Performance
ANN	artificial neural network
APB	acid-producing bacteria
API	American Petroleum Institute
ASCE	American Society of Civil Engineers
AST2	Type 2 aluminized steel
ATP	adenosine triphosphate
AWWA	American Water Works Association
BMP	best management practices
BMPDB	Best Management Practices Database
CEC	cation exchange capacity
C.F.R.	Code of Federal Regulations
CIPS	close interval potential survey
CP	cathodic protection
CSE	copper–copper sulfate reference electrode
DC	direct current
DIGGS	Data Interchange for Geotechnical and Geoenvironmental Specialists
DNA	deoxyribonucleic acid
DSS	decision support system
EIC	environmentally induced cracking
EIS	electrochemical impedance spectroscopy

EMI   electromagnetic induction
ER   electrical resistance
ERC   Engineering Research Centers

FBE   fusion-bond epoxy
FHWA   Federal Highway Administration
FISH   fluorescence in situ hybridization

GDP   gross domestic product
GSI   Geosynthetic Institute

ICCP   impressed current cathodic protection
ILI   in-line inspection
IRB   iron-reducing bacteria

LPR   linear polarization resistance

MCEER   Multidisciplinary Center for Earthquake Engineering Research
MFL   magnetic flux leakage
MIC   microbially influenced corrosion
ML   machine learning
MSE   mechanically stabilized earth
MURI   Multidisciplinary University Research Initiative

NAS   National Academy of Sciences
NBS   National Bureau of Standards
NSF   National Science Foundation

PCR   polymerase chain reaction
PDP   potentiodynamic polarization
PE   polyethylene
PVC   polyvinyl chloride

RNA   ribonucleic acid

SCC   stress corrosion cracking
SIP   spectral induced polarization
SOB   sulfur-oxidizing bacteria
SP   standard practice
SPP   sulfide-producing prokaryote
SRB   sulfate-reducing bacteria

USGS   U.S. Geological Survey